T0386834

Professor SHINJI TAKAHASHI, M. D.
Department of Radiology, Nagoya University, School of Medicine,
Nagoya/Japan

Professor SADAYUKI SAKUMA, M. D.
Department of Radiology, Nagoya City University, School of Medicine,
Nagoya/Japan

ISBN-13: 978-3-642-66122-8 e-ISBN-13: 978-3-642-66120-4
DOI: 10.1007/978-3-642-66120-4

Library of Congress Cataloging in Publication Data. Takahashi, Shinji, 1912- Magnification radiography. Includes bibliographical references and index. 1. Radiographic magnification. 2. Diagnosis, Radioscopic. I. Sakuma, Sadayuki, 1931-joint author. II. Title. [DNLM: 1. Radiographic magnification. WN100 T136m] RC78.T116 616.07'572 75-9778.



Softcover reprint of the hardcover 1st edition 1975



Typesetting, printing, and bookbinding by Universitätsdruckerei H. Stürtz AG, Würzburg

# Preface

If the early stages of a disease begin with the involvement of a small area of cells or tissue, the early diagnosis of pathologic changes by means of radiography should concentrate first on the detection of such minute changes. The ideal solution would be to produce X-ray images of findings much finer than those observable by the naked eye, and herein lies a new field of research that is believed to be worth developing.

The introduction of a 0.3 mm focal-spot rotating-anode tube about 25 years ago opened the way to the clinical application of magnification radiography. Due to the postwar economic situation, we were unable to import this type of X-ray tube, but we believed in the importance of magnification radiography in X-ray diagnosis, and in 1952 we produced an X-ray tube with a 0.15 mm focal spot by reconstructing an existing fixed-anode tube. This X-ray tube has been improved step by step, so that tubes with focal spots of 0.1 mm or 0.05 mm are now available in Japan. Thus it has become possible to obtain 4 to 6 × magnification images of minute lesions that could not be imaged by normal roentgenography.

In other countries an X-ray tube with 0.1 mm focal spot has also been manufactured very recently, and the clinical value of magnification radiography has been confirmed, especially in angiography. In view of this worldwide trend toward the increased application of magnification radiography, it was considered timely to publish a book describing our experiences in detail.

The main drawback of magnification radiography has always been that the skin dose increases with the magnification ratio. In reviewing the applicability as well as the limitations of magnification radiography, we have especially kept in mind the question of patient dose reduction. Our results are discribed in detail in this book.

Preparation of the manuscript was made possible by the help of Mr. T. Tomiyoshi, who carried out the magnification radiography, Mr. H. Maekoshi, who drew the illustrations, Mr. S. Fujita, who prepared the photographs, and Miss T. Nishikawa, who arranged and typed the original manuscript. Dr. T. OYAMA undertook the labor of translating our manuscript from Japanese into English.

We wish to take this opportunity to express our deep appreciation to all these persons, as well as to former and present staff of our Departments.

Nagoya, July 1975

SHINJI TAKAHASHI
SADAYUKI SAKUMA

# Contents

# Introduction

There are two ways of obtaining a magnified image on roentgenograms. One is the method of first taking a clear and sharp roentgenogram and having the image enlarged optically, or indirectly. As the applicability of this method to clinical medicine is limited, indirect magnification roentgenography is not described here.

The other method is based on the object being placed midway between an X-ray tube with a small focal spot and the X-ray film. As the X-rays diverge, a magnified image is obtained directly on the X-ray film. This method is simple in technique and provides rich and significant findings for diagnosis; its details are described in this book.

There have been many terms for this type of roentgenography conducted by direct magnification, such as microradiography (VALLEBONA [125], TAKAHASHI [102]), enlargement radiography (ALLEN and ALLEN [5], KOMIYAMA [44], TAKAHASHI [104]), magnification radiography (BOOKSTEIN [12], FRIEDMAN [29]), direct macroradiography (AYAKAWA [7]), macroroentgenography (IKEDA [37], IMAGUNBAI [38]), macroradiography (SAMUEL [82]), radiographic enlargement (FLETCHER and ROWLEY [28]), direct X-ray enlargement (HACKER [33]), direct enlargement radiography (MATSUDA [50]), X-ray image magnification (NEMET [55]), radiologische Vergrößerungstechnik (ADERHOLD and SEIFERT [3], VAN DER PLAAT [61]), förderliche Röntgenvergrößerung (ADERHOLD [4], FERRANT [27]), direkte Vergrößerungsaufnahme (SAKUMA [73]), images radiologiques agrandies directement (SVOBODA [100]), and others.

In the present book this type of roentgenography is described as direct magnification radiography, shortened to magnification radiography. When the magnification ratio is mentioned, we refer to direct radiography of $\alpha \times$ magnification or, for short, $\alpha \times$ magnification radiography.

First we make a general review of the field, and explain in what respects the X-ray tube differs from conventional types. Next, we deal with the need to know in interpreting the X-ray pictures the special features that are not seen in the normal roentgenogram. In view of the need to protect the patient from undue radiation, reduction of the skin dose is also discussed. Next, the type of apparatus and the techniques necessary for satisfactory magnification radiography are described in detail.

Finally, we discuss the actual interpretation of the finer parts of the body as revealed by magnification radiography, on the basis of a comparison of X-ray images taken clinically by both normal roentgenography and magnification radiography.

# *Chapter I.* Principles of Magnification Radiography

## A. Magnification Radiography in General

In magnification radiography, the distance between the X-ray tube and the object or the film, and the size of the focal spot of the X-ray tube are in the following relationship:

If the diameter of the focal spot of the X-ray tube is $F$, the size of the object to be radiographed $AB$, the distance between focal spot and object $m$, the distance between object and film $n$, and the area of blurring caused by the penumbra at both ends of the X-ray image $H$, then $H=\frac{n}{m}F$ (Fig. 1A). With increasing $H$ the image loses its sharpness and a very small object is blurred out. To prevent this, it is necessary to keep $H$ as small as possible.

By using a tube with a very small focal spot a magnified image $A'B'$ is obtained directly on the film, as shown in Fig. 1B. The relationship between image and object will be

$$A'B'=\frac{m+n}{m}AB=\left(1+\frac{n}{m}\right)AB.$$

If now the magnification ratio is taken to be $\alpha$:

$$\alpha=\frac{A'B'}{AB}=1+\frac{n}{m}.$$

In this type of roentgenography intensifying screens are used, and these produce blurring. However, even extremely small

Fig. 1A–C. Factors related to magnification radiography. A) Relationship between size of penumbra and size of focal spot. B) Relationship of magnification ratio between focal spot-object distance ($m$) and object-film distance ($n$). C) *Left:* Blurring of image of small object due to screens as used in normal roentgenography: small object is almost obliterated due to blur. *Right:* Blurring of image of small object in magnification radiography: small object is imaged. $F$ Focal spot; $H$ Penumbra; $AB$ Object; $A'B'$ Image; $S$ Screen

objects are magnified and imaged, provided the focal spot size of the tube is sufficiently small, because their images become larger than the blurring or mottling produced by the screens. From the aspect of detection of the lesion, therefore, the use of intensifying screens does not interfere very much with the examination of the fine structure of the image (Fig. 1 C).

No grid is used in magnification radiography. As the distance between object and film becomes greater, the Groedel or air-gap effect arises. Fewer scattered rays reach the film and the contrast of the picture is improved. If a tube with an appropriately small focal spot is used for magnification radiography, it is possible to obtain X-ray images of minute tissues and pathologic foci of 0.2 mm diameter, and down to 0.025 mm.

Although visualization of the object depends greatly on the contrast of the image, a very small object is essentially not visible on the normal roentgenogram. There are many cases where it is impossible to make a correct diagnosis unless magnification radiography is performed.

Magnification radiography is a special roentgenographic technique. In X-ray examination, a normal roentgenogram is first made together with a general examination of the patient. Sometimes, although no finding can be seen on the normal roentgenogram, certain diseases are suspected from the history or the current status of the patient, for instance, in early silicosis, where nodulations or slight fibrosis are not imaged and it is difficult to make a correct diagnosis. With magnification radiography, the smallest lesion can be found and a correct diagnosis can be made.

There is also the case, where an abnormal shadow with pathologic change is imaged on the normal roentgenogram, but the fine parts involved are not recognizable. Magnification radiography conducted as an aid to differential diagnosis, for instance, as between carcinoma and scar in angiographic examinations, will be useful for obtaining detailed findings.

The advantages of magnification radiography also lie in its ability to allow the image of minute lesions to be viewed easily, thus facilitating diagnosis. Existing magnification radiography by means of a 0.3 mm focal-spot tube has contributed to such diagnosis, and the smaller the focal spot size, the greater will be the advantages of magnification radiography.

Even when the normal roentgenogram contains sufficient detail for a diagnosis to be made, there are cases where it is advisable to confirm the diagnosis by the use of additional techniques of greater accuracy. For example, where the prognosis is considered to be poor, some hope for the better may be obtainable if a more accurate diagnosis can be made from more detailed findings.

A single small lesion, which may influence the diagnosis as a whole but is recognized with difficulty or overlooked, can be the subject of examination by magnification, as it will then become easy to detect. In such a case magnification radiography will be an appropriate means.

Besides the above, the improvement of the sharpness of images by means of magnification photofluorography is considered valuable. In normal roentgenography image enlargement merely produces lack of sharpness.

Essentially, however, the merit of magnification radiography lies in its ability to reveal new findings based on the minute X-ray images which cannot be seen by normal roentgenography. This characteristic should be the primary objective of magnification radiography.

A warning should, however, be given: the fact that this improvement in technique providing many more findings than by normal roentgenography does not in itself always lead to the establishment of a different diagnosis.

Current diagnostic radiography is based on the interpretation of findings obtained not by magnification but only by the naked eye. Magnification radiography should open up a new radiodiagnostic field and promote the establishment of a much more accurate diagnosis. The technique is at present considered to be in the development stage.

## B. Historical Review

To carry out magnification radiography, an X-ray tube of small focal spot has to be prepared. The first attempt to prepare such a tube was made by VALLEBONA [125], who placed a 3-mm thick brass plate with a pinhole of diameter 0.5 to 1.0 mm at the radiation mouth of an X-ray tube. By this method he was able to obtain X-ray images of about 2× magnification of the femur, carpal bones and/or sella turcica in the dry state of bone.

This method was also attempted by ZIMMER [140] and TAKAHASHI [103], but it failed to become practical.

BURGER *et al.* [18] and COMBEE and BOTDEN [21] prepared for the first time a focal spot of 0.3×0.3 mm in the rotating-anode tube. Following the introduction of this minute focal-spot tube, magnification radiography began to be the subject of discussion in various countries. FLETCHER and ROWLEY [28] stated that magnification radiography might be inferior to indirect or optical magnification radiography. GILARDONI [31] commented on the basis of various aspects of radiological physics that indirect magnification radiography was better than the positive use of a 0.3 mm focal spot tube. BÜCHNER [15, 16, 17] and TAKAHASHI [104] stated that a 2× magnified X-ray image obtained by roentgenography with a 0.3 mm focal spot tube is inferior in the formation of an image, both theoretically and practically, to an X-ray image obtained by normal roentgenography using intensifying screens.

ALLEN and ALLEN [5] stated that for magnification radiography a focal spot substantially less than 1.0 mm is essential, and that this would greatly improve the possibility of resolution of fine structures.

VAN DER PLAATS [61] published a paper on the technical aspects of magnification radiographic procedure using a 0.3 mm focal-spot tube. ZIMMER [140] stated that magnification radiography was a help in the search for minute pathologic changes and confirmed [141] the usefulness of magnification radiography in facilitating the observation of findings in more detail, although no new findings were seen in magnification radiography. WERNER and BADER [134] took a similar view.

WOOD [135] stated that a 0.3 mm focal-spot tube was useful, especially for examination of the bone. SAMUEL [82] was of the same opinion regarding diagnosis of the cranial bones of the ear, nose, and throat.

SEYSS [90] using a tube with a focal spot of 0.3 mm, obtained more detailed structure, particularly of the fine, diffuse fibrosis of old age.

FERRANT [27] had a favorable opinion of magnification radiography. ABEL [1] considered magnification radiography performed by 0.3 mm focal-spot provided a sharp magnified image. BAKER *et al.* [11] stated that magnification radiography was of value in the diagnosis of the spine in forensic medicine. Lung markings were studied by magnification radiography (YOSHIDA [138]), and SHINOZAKI and KOMIYAMA [91] tried magnification radiography in 2.5× magnification for early silicosis.

SVOBODA [99, 100] proved the usefulness of magnification radiography for localized replacement of the bone in hemophilia of the joint, and RANDALL [62] for localized stenosis of the coronary artery.

KEATS [43] reported that in the lungs of children the magnification radiogram presented better findings than the normal roentgenogram, while TAKARO [122] noted that magnification radiography not only facilitated observation when applied to angiography but also enabled the image of the glomeruli to be observed.

TAKARO and SCOTT [121] reported several excellent magnified angiograms of renal, cerebral, abdominal, pulmonary, and peripheral vessels. They considered that, for magnification radiography of the heart and kidneys, the capacity of the tube should be made higher.

Even when a 0.3 mm focal spot tube is used, the resolution at 2× magnification is better than that of a normal roentgenogram because of the use of the small focal spot and the air-gap effect (BOOKSTEIN and VOEGELI [12]). A short distance between target, film and grid provides good magnification but with a high skin dose (BOOKSTEIN and POWELL [14]). As for the image quality of a magnification radiograph of a moving object, ANGERSTEIN and STARGARDT [6] wrote a paper on modulation transfer functions.

HACKER [33] recognized the applicability of direct magnification radiography to cerebral angiography, and LLOYD *et al.* [48] applied this technique to the demonstration of the lacrimal drainage system. Direct radiographic magnification of laryngotracheal lesions in infants and children was carried out with good results by LALLEMAND *et al.* [47] and CALENOFF and NORFRAY [19] used a 0.3 mm focal-spot tube for renal osteodystrophy in hemodialyzed patients. Direct magnification intravenous pyelography for the evaluation of medullary sponge kidney was reported by HAYT *et al.*

Summarizing the above, it can be said that those who assert that this method is of value are radiologists who base their conclusions on clinical experience, while those who believe that 2× magnification radiography using a tube with a focal spot of 0.3 mm may not help in finding new pathologic foci include a few radiologists and many physicists who base their views mostly on theoretical considerations.

The availability of an X-ray tube with a focal spot less than 0.3 mm in diameter would enable the usefulness of magnification radiography to be evaluated.

SEIFERT [89] reported on an experimentally produced tube with a focal spot adjustable between 1 mm and 0.03 mm ± 0.005 mm. This tube was so constructed that X-rays were emitted from a part of the tube wall which was made to be the target of the anode. The output was from 5 mA to 0.5 mA. The maximum magnification ratio was 11 times. ADERHOLD [4] and SEIFERT [89] radiographed various parts of the body with this tube, and it was hoped that more than 2× magnification radiography would provide much more precise findings than normal roentgenography.

In Japan, a radiation source with a 0.2 mm spot source of $^{35}S$ was experimentally prepared and magnification radiography was attempted with the 0.17 MeV electron rays emitted from $^{35}S$ (TAKAHASHI and KOMIYAMA [103]). However, it was concluded that this method could not be applied clinically. Next, 150 kV X-radiation emitted

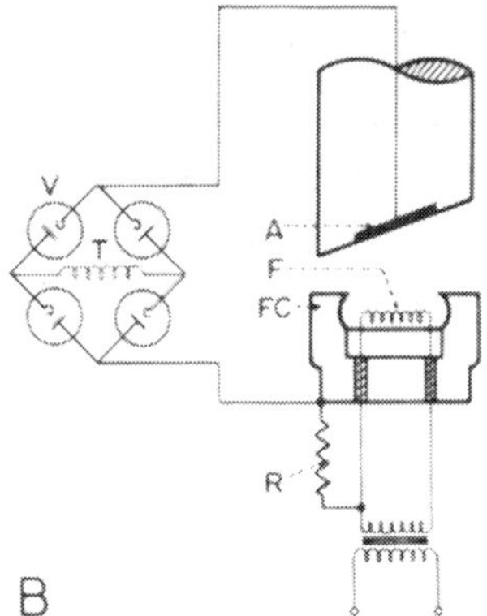


Fig. 2A and B. A) Grid-biased focal-spot tube reconstructed from the usual fixed-anode tube (Toshiba Sealex). Lead wire (arrow) from the focusing cup. B) Circuit diagram of grid-biased focal-spot tube. *A* Target; *F* Heating filament; *FC* Focusing cup; *R* Resistance (variable). (Presented at the Hirosaki Meeting of Medicine, Hirosaki, 19 December 1953)

through a pinhole, as in VALLEBONA'S device [125], was tried for experimental purposes, but this also proved to be impractical (TAKAHASHI and KOMIYAMA [103]).

One year before SEIFERT'S [89] paper on his new X-ray tube, TAKAHASHI used a fixed-anode tube with a lead wire emerging from the focusing cup and isolated the wire from the socket of the tube (Sealex, Toshiba). The terminal end of the lead wire was taken out of the tube to be free from the heating filament (Fig. 2A and B). In the circuit of focusing cup and heating filament a 70 kΩ resistance was inserted, so that the difference in potential between focusing cup and heating filament became 1400 volts. This difference in potential produced a bias phenomenon in the electron beams from the heating filament and enabled the size of the original focal spot to be reduced from 5 × 5 mm to 0.15 mm. This tube was applied to magnification radiography of the elbow and chest of an adult human. The X-ray tube was named the biased focal-spot tube (TAKAHASHI, 17th Hirosaki Meeting of Medicine, 19 Dec. 1953; TAKAHASHI and KOMIYAMA [105]; TAKAHASHI [104]; TAKAHASHI *et al.* [107, 108]). In order to estimate the size of the focal spot, an attempt was made to calculate it from the resolving power (KOMIYAMA [44]).

With this tube having a focal spot less than 0.3 mm in size, observation of the fine trabeculation of the bone (MATSUDA [50]), of the pulmonary markings in 2 to 3 × magnification (YOSHIDA [138]), and of pulmonary tuberculosis (KOMIYAMA *et al.* [46]) first became possible clinically. Theoretical and experimental studies made it clear that in magnification radiography the use of medium-speed screens (MS) is appropriate (KOMIYAMA [45]). By means of this fixed-anode tube experimental axial transverse cross-section images of the object were taken by TAKAHASHI *et al.* [106].

For clinical practice, a rotating-anode tube is obviously better than a fixed-anode tube. Therefore a very finely biased focal-spot rotating-anode tube was produced with the cooperation of Toshiba Co.

This type of tube, however, has an output of only about 2 mAs (TAKAHASHI *et al.* [107, 108]). The size of the focal spot of this tube was calculated to be 0.055 mm (TAKAHASHI and WATANABE [110]). The use of this ultra-fine focal-spot tube allowed a test object of wires 0.025 mm in diameter, arranged parallel to each other and 0.025 mm apart, to be resolved in the magnification radiograph at 10 × magnification. The characteristics of this tube were determined and reported (SAKUMA *et al.* [78]) and several experiments were made to determine the specific effects of magnification radiography (TAKAHASHI and YOSHIDA [109]; YOSHIDA [139]).

The experiments showed that in magnification radiography an appropriate contrast medium is useful for good imaging (WATANABE [127]).

With this tube, a further experimental study on specific effects in magnification radiography was made of pulmonary markings (SHINOZAKI *et al.* [92]) and of bone (YATO [136, 137]). In the orthopedic field, magnification radiography was also useful (SUGIURA [98]).

In 1960 SAKUMA [67] conducted a comparative study of lung markings in normal adults by normal roentgenography and 2 and 4× magnification radiography. The findings of the minute marking areas were enhanced in 4× more than in 2× magnification radiography. The appearance of numerous small ring shadows was especially noticeable.

On examining a case of early pulmonary silicosis by 4× magnification radiography, a diagnosis could be established earlier than by the current use of normal roentgenography and 2× magnification radiography. There was a marked increase in the findings of punctiform shadows of around 1 to 2 mm diameter and small ring shadows of 2 to 3 mm diameter (TAKAHASHI *et al.* [112]; SUGIE [97]; ONO [58]).

In pulmonary tuberculosis, 4× magnification radiography was not of much value for exudative or infiltrative lesions but gave better results for proliferative lesions, the sclerotic stage, calcification, and pathological changes in the bronchi than normal roentgenography (TOKUNAGA [124]).

SAKUMA and KOGA [68] found that the villi of the small intestine of adults were visible only on the 4× magnification radiograph. These villi were not imaged by normal roentgenography or existing magnification radiography. In lymphography with 4× magnification the linear shadows of the minute medullary cords increased while after 20 hours the punctiform shadows of retention of contrast media in the sinuses were seen to be more numerous than with normal lymphography. Lymphography by serial magnification radiography revealed the fine structure of the peripheral and central sinuses (SAKUMA *et al.* [70, 71, 74]).

After lymphography of the foot by the Kimmoth method the production of an oily embolism in the lung was seen much more clearly and in greater detail when compared with the findings by normal roentgenography (SAKUMA *et al.* [70, 71, 73]).

The application of 4× magnification radiography is of particular value in angiography (TAKAHASHI *et al.* [113]). Some techniques were devised by KANEKO [40] and applied to the renal artery (KANEKO and SAKUMA [42]). Tumor-specific findings were much better seen on the magnification angiogram than on the normal angiogram (TAKAHASHI *et al.* [114]; SAKUMA *et al.* [77]). Neuroradiological application was reported for the first time for the vessels of the orbit (SAKUMA [69]) and later for the brain (SASAKI *et al.* [85]).

The punctiform shadows that are seen in the cortex after 4× magnification radiography of the renal blood vessels were found to be the glomeruli in human adults (KANEKO [41]). Again, when 4× magnification radiography was made of selected pulmonary arteries, the terminal arteries and part of the arterioles became visible (SAKUMA and MIURA [75]).

With magnification radiography of the bronchial artery in pulmonary cancer, the state of the vascular wall, the brushlike shadow of the tumor, etc. were much more clearly seen than in normal bronchial arteriography (SATO [87]; SASAKI [86]).

Further, in arteritis, 4× magnification radiography revealed the minute changes in the aorta, medium and small arteries. This method was also applied to uterine tumor (TAKAHASHI and NAGATA [101]). Hypotonic gastroduodenal 4× magnification radiography revealed the area gastrica in more than half of the patients examined and was believed to be of value in the diagnosis of early gastric cancer (SAKUMA *et al.* [76]).

A 20× magnification radiograph was taken with a 0.05 mm focal-spot tube. The image appeared blurred at first glance. However, it proved to be rich in information, and the diagnosis was enhanced when the radiograph was observed after reduction optically into 6× magnification (SAKUMA *et al.* [79]).

4× magnification lymphography was superior to normal lymphography in the diagnosis of tumor and metastasis (IMAGUNBAI [38]). Both the theory and the clinical applications of 4× magnification radiography have been summarized (TAKAHASHI *et al.* [115]; TAKAHASHI *et al.* [118]).

Favorable results were obtained by 4 or 6× magnification radiography with a radiation field of 2×2 cm applied to the lung and bronchi and in lymphography (SAKUMA *et al.* [80]) with a stereographic technique.

Studies were made where high magnification radiographs and normal roentgenograms of the same tissue or lesions were compared. The conclusion was that the former were superior in imaging the minute lesions. Magnification radiography has the specific merit of leading to the correct diagnosis as compared with normal roentgenography or axial transverse tomography (SAKUMA *et al.* [72]).

From the viewpoint of image quality, the optimal magnification was experimentally determined by means of the modulation transfer function (MTF) or response function (AYAKAWA *et al.* [7]).

First, a study was made of the MTF of magnification radiography with special reference to the tube focus (OKUMURA *et al.* [57]), followed by a study of the MTF of the film and intensifying screen, using a 1.5 mm, 0.3 mm, 0.1 mm or 0.05 mm focal spot tube (AYAKAWA [8, 9]).

A problem in magnification radiography is that the human body emits scattered rays and this affects image quality. This problem was studied next. Even where the thickest part of the body was being examined, more information was obtained with high-magnification radiography conducted with various small focal-spot tubes than with normal roentgenography (AYAKAWA and SAKUMA [9]).

There is now no doubt that magnification radiography carried out with a 0.05 mm focal-spot tube is more useful than normal roentgenography when applied to nonmoving parts of the body. However, SAKUMA [81] stated that for a moving organ in a thick part of the body, magnification radiography with a larger focal spot, i.e. 0.3 mm, can sometimes be better when the grade of movement is large because of its high output.

Recent publications report successful applications of the 0.05 mm focal-spot tube to magnification stereoradiography (IKEDA [37]) and to the diagnosis of gout (ISHIGAKI [39]). TAMAKI [123] read a paper on the clinical application of dental magnification radiography at the Third International Congress of Maxillofacial Radiology.

An X-ray tube with a focal spot 0.1 mm in size was also found useful for the thick part of the body by WATANABE [128].

In countries other than Japan, an X-ray tube having a focal spot smaller than 0.3 mm had not been produced so far, except by

ADERHOLD. In Japan, we have experience with both 0.05 mm and 0.1 mm focal-spot tubes.

As an X-ray tube with a focal spot smaller than 0.3 mm became available in Europe and U.S.A. The worldwide clinical application of the tube has been becoming more frequent, especially in the field of angiography and neuroradiography.

By reducing the target angle to 10°, GREENSPAN *et al.* [32] succeeded in narrowing the effective focal spot to 0.26 mm and reported on its value in clinical magnification radiography. According to ABLOW *et al.* [2], magnification radiography of the newborn chest by means of this tube was employed in a study of more than 150 infants. The satisfactory results obtained were endorsed by CREMIN [22].

DÜNISCH *et al.* [26] successfully prepared a tube with a focal spot of 0.1 mm. PFEILER and THEIL [60] discussed the pseudoresolution of lymph vessels. HOLLANDER [36] experimented with a 0.05 mm focal-spot tube produced for trial purposes by Picker Co., and reported that minute changes in lesions could be observed with 4 × magnification radiography of the cranial blood vessels.

According to SAGEL and GREENSPAN [88], cases examined by their X-ray tube showed tiny arteriovenous malformations in the lung, as demonstrated by direct magnification pulmonary angiography.

SIDAWAY [93] performed magnification radiography of the small vessels of the kidney by means of this tube. Interlobar vessels beyond the power of definition were brought up to within definition.

In Germany, magnification radiography has been applied very actively in the field of cerebral angiography with a 0.3 mm or 0.1 mm focal spot tube, by WENDE *et al.* [130, 133]. Better definition and visualization were achieved for the veins in vertebral angiography by WACKENHEIM *et al.* [126]. For cerebral tumor, although isotope scintigraphy should not be omitted as the pre-examination, serial magnification angiography alone was able to differentiate the tumor from the infarct and it was possible to get minute important findings (WENDE *et al.* [132]).

The magnification technique has also been applied to orbital disease (WENDE and NAKAYAMA [131]). The tentorial artery was visualized by means of magnification cerebroangiography after operative treatment of congenital arteriosinus fistula (NAKAYAMA *et al.* [54]).

Experimentally, the conclusion was reached that, while a microfocal-spot X-ray tube has its place in radiography for the improvement of resolution, it is best used in the conventional way with a magnification factor as close to unity as possible, except where X-ray tubes with foci of 0.1 to 0.2 mm in size are available (RAO and CLARK [63]).

Magnification radiography with a tube of 0.3 mm focal spot has proved to be of value clinically, because it enables new disease foci to be observed. Recently the value of the technique has been enhanced because the magnification rate can now be increased by the use of a still smaller focal spot, i.e. 0.28 mm, 0.1 mm, or 0.05 mm.

DOI [23, 25] has published useful assessments of the image quality of normal roentgenography and magnification radiography. Small vessels will be recognized when magnification radiography is carried out with the appropriate focal spot. According to DOI's computer program simulating a small blood vessel of 0.1 mm diameter, the image contrast increases almost four times when a magnification of 20 × is used with an 0.1 mm focal spot (DOI [24]). SANDOR and ADAMS [83] described their computer model which provided rapid assessment of the expected image amplitude of a given blood vessel diameter, magnification image distortion, and the effect of a contrast agent on image quality and detectability.

According to MILNE [51], with magnification radiography it is possible to visualize a blood vessel as small as 15 µm in diameter. He emphasized the importance of the effect of the focal spot, which emit X-rays with Gaussian distribution. SANDOR *et al.* [84] stated recently that magnification radiography can visualize findings not imaged on a normal roentgenogram. This effect is partially due to magnification.

Magnification radiography applied to small vessels imaging is estimated valuable by HILAL, BOOKSTEIN, EISENBERG and others [35].

## C. X-Ray Tube

In magnification radiography it is necessary to take extremely clear, sharp pictures. The result chiefly depends on the combined effect of size of focal spot and exposure time. Knowledge of the X-ray tube to be used for this purpose is the basis for successful performance of magnification radiography.

### 1. Focal Spot

The essential tool of magnification radiography is the focal spot of the X-ray tube. The tube used currently for magnification radiography does not differ in appearance from the usual rotating-anode X-ray tube yet, unlike the existing X-ray tube, its focal spot is in comparison extremely small (Fig. 3).

With a given electric potential between the focusing cup, heating filament, and the tube current, the electron beam takes on a definite size at the target surface.

In the usual X-ray tube for normal roentgenography the potentials between the filament and the cup are caused to be equal by making a short circuit between them. For

Fig. 3A–F. Focal spot of X-ray tube in 20× magnification radiography. Pinhole camera with 0.03-mm diameter aperture. A) 0.3 mm focal spot; B) 0.1 mm focal spot (grid-biased); C) 0.05 mm focal spot (grid-biased); D) 0.1 mm focal spot (unbiased); E) 0.1 mm focal spot (grid-biased for D); F) 0.1 mm focal spot (non grid-biased focal spot tube; subfocus image is seen)

2 × magnification radiography, the existing X-ray tube having a focal spot of 0.3 mm in size has also been constructed on this principle. If an X-ray tube is manufactured on the same principle for higher than 2 × magnification radiography, off-focus radiation tends to occur.

When radiography of the focal spot is performed with a pinhole camera with an aperture of 0.03 mm diameter, a "wing" or "subfocus" of off-focus radiation is imaged. Off-focus radiation produces fog on the roentgenogram. This phenomenon means that the thin parts of the body (less than 10 cm thick) appear in poor contrast, so that a correct diagnosis is established with difficulty. When the object is thick, the bad effects of subfocus on X-ray image formation are relatively minor due to the additive effect of the scattered rays emitted from the body to the off-focus radiation. Therefore only in such a case can the X-ray tube be applied to magnification radiography. The X-ray tube with a subfocus is of limited applicability, although its output is rather high.

Before the size of the electron beam which produces a very fine focal spot can be regulated at will, the electric potential of the focusing cup has to be appropriately changed. This operation induces bias of the electron beam according to the voltage range, and the required size of beam to produce the small focal spot can be obtained. Such a tube is termed the grid-biased focal-spot tube.

The grid-biased focal-spot tube is a linear focal spot unless bias arises in the electron beam (Fig. 3D), (TAKAHASHI [107, 108]). The electron density at the focal spot of the target is not uniform and usually consists of two parallel lines. If the voltages at the cup and filament of this tube are correctly adjusted, the two lines will gradually approach each other and overlap in homogeneous density, due to the bias of the electron beam; they also become shorter and form a small beam. If the voltage is raised still further, the overlapping lines will cross over and again form two lines in the image. The required voltage is called the grid-bias voltage and is used for obtaining the minimal focal spot. The voltage necessary for obtaining a focal spot of homogeneous electron density is called optimal bias voltage.

A transformer is usually used for supplying grid-bias voltage. For a certain tube voltage, there is a definite optimal voltage. When the tube voltage is below a certain figure, for instance 120 kV, the minimal focal spot enlarges even when the bias voltage remains fixed, so that it will be necessary to raise the bias voltage. This operation becomes tedious if an adjustment has to be made every time. Therefore, a self-bias apparatus, which always ensures optimal bias voltage irrespective of the variation in tube voltage, has now become available. The size of the focal spot depends on the tube current (RAO and SOONG [65]). The self-bias apparatus is certainly of value in the clinic, but it is advisable to measure the size of the focal spot from time to time and to make regular checks to ensure that the grid-bias phenomenon is working correctly.

The commercially available X-ray tubes of focal spot 0.05 mm by TOSHIBA, those of 0.1 mm by TOSHIBA, HITACHI and SIEMENS and that of less than 0.3 mm by MACHLETT are manufactured on this principle. A tube with a very small focal spot is free from subfocus and can be used for magnification radiography of every part of the body.

There is, however, another procedure for producing an effective small focal spot. When the target angle is $\alpha$, the actual focal spot $a$ is reduced in size when observed perpendicularly from the anode-cathode axis of the X-ray tube. It then becomes the effective focal spot $b$ and

$$b = a \sin \alpha.$$

When $\alpha$ is a small angle, $\sin \alpha$ is also small, resulting in an effective focal spot of small size. Tubes with very small focal spots have different target angles according to the manufacturer's policy.

The size of the effective focal spot imaged on the X-ray film by means of a pinhole camera is usually different in two directions, i.e. perpendicular and parallel to the anode-cathode axis. According to IEC 336 (International Electrotechnical Commission, Publication 336, 1970) the tolerance of the size of the focal spot is $+50\%$ when the nominated size of the focal spot is less than 0.8 mm.

Because the focal spot differs in size in two directions that are perpendicular to each other, it causes a difference in sharpness on the roentgenogram. Just as the effective size of the focal spot differs in accordance with target angle, the intensity of the X-rays can also differ with the direction of the central X-rays to the target surface. This becomes a problem when the target angle $\alpha$ is too small (Komiyama [44]). Therefore, when magnification radiography is conducted with large beam size and too short a distance between focal spot and patient, due caution is necessary.

## 2. Measurement of Focal Spot Size

In order to confirm the small size of the focal spot, a measurement should be performed periodically, both directly by the pinhole-camera method, and indirectly by the resolving-power or MTF-curve method.

### a) Pinhole-Camera Method

Focal spot size is measured directly by means of an X-ray pinhole camera, which must have a pinhole diameter at least one tenth that of the focal spot. The pinhole diameter recommended by ICRU Report 10f 1962 is 0.03 mm for a focal spot of less than 1.0 mm diameter. Hence, for a tube with a 0.1-mm or 0.05-mm focal spot an X-ray pinhole camera of pinhole diameter 0.005 to 0.01 mm is appropriate. To prepare a diaphragm with such a small pinhole diameter poses many problems from the point of view of technique and material quality, not to mention the difficulty of obtaining a sufficiently dense image.

Measurement with an X-ray pinhole camera of 0.03-mm aperture is, however, adequate for a focal spot of 0.1 to 0.05 mm. The pinhole camera method is of particular value when the shape of the focal spot is markedly distorted, i.e. when subfocus and/or wing is present beside the main focal spot (Fig. 3). With the pinhole camera the minimum distance from focal spot to pinhole is 10 cm; when X-rays of kVp are used the magnification ratio is usually at least 10 times better. Screens are not used. The other factors are as recommended in IEC publication 336 (1970). Observation of the image of the ultra-small focal spot will be easier if the magnification ratio is made as large as possible by increasing the distance between pinhole and film. Even with a magnification ratio of 20, the size of the focal spot image will be only 1.0 mm or 2.0 mm when a 0.05 mm or 0.1 mm focal-spot tube is used. The focal spot, pinhole, and the center of the film are adjusted on the central X-ray. To ensure an accurate magnification ratio, the location of the pinhole is precisely measured and moved when the location of the focal spot becomes clear by triangulation. It is advisable to

Fig. 4. Density curve of focal spot. Scanning is done through the center of the focal-spot image of Fig. 3C

conduct roentgenography when the densest part of the image of the focal spot is between 0.8 and 1.2 in density. The density curve scanned at the center of the focal spot image of Fig. 3C is shown in Fig. 4. This procedure enables the shape and size of the image of the focal spot to be easily and concretely obtained, although not always with the utmost accuracy (SAKUMA *et al.* [78]).

b) Estimation with Resolving Power

It is possible to determine focal spot size by means of a test object (cf. Chap. I. F. 1.) made up of groups of three metal wires of various diameters (KOMIYAMA [44]). If the diameter of the focal spot does not exceed twice the intervals between the wires, it will be possible to resolve the test object of parallel metal wires arranged at intervals equal to the wire diameter.

If now we take a magnification roentgenogram of the test object consisting of wires of diameter $b$ placed parallel at intervals of $b$, and the distance from the focal spot is $m$, the distance between the object and the film is $n$, and the size of the effective focal spot is $a$, $n$ will be negligibly small in normal clinical roentgenography, so that there will be only a slight blur due to the penumbra (Fig. 5).

Fig. 5. Diagram of resolution mechanism of test object composed of wires of diameter $b$ arranged parallel at intervals of $b$, radiographed by X-ray tube with focal spot of diameter $a$. $m$ Focal spot-test object distance. $n_1$ Distance between test object and point of minimum umbra where contrast of object becomes lower. $n_2$ Distance between test object and point where test object is imaged homogeneously. $n_3$ Location where pseudoresolution arises

When $a > 2b$, and $n$ is much larger than $m$, the image of the test object becomes more blurred due to the increase of the size of the penumbra and reduction of the umbra, then becomes homogeneous, and finally pseudoresolution occurs.

When $b < a \leq 2b$, the test object is resolved only when the distance between the focal spot and the film is small. If that distance is increased, the image loses contrast and homogeneity, so causing the structure of the test object to disappear.

When $a = b$, the umbra retains its original size but the penumbra increases, so that when the diameter of the wire is increased as well as the intervals, the structure of the test object can still be seen.

When $a < b$, both umbra and penumbra increase in size, and the wires of the test object can be recognized on a roentgenogram of high magnification. If the focal spot is smaller than twice the diameter of the wire, the wire is resolved no matter how high magnification is used.

The size of the focal spot can be estimated from the above statements as follows:

If the test object of 0.025-mm wires is resolved on the magnification radiogram at a low magnification ratio but barely resolved at a high magnification ratio, we can conclude that the focal spot is equal to or smaller than 0.05 mm. If a 0.032-mm test object is clearly imaged at the same magnification ratio, the focal spot size will be larger than 0.032 mm. Therefore

$$0.05\ \text{mm} \geq \text{focal spot size} > 0.032\ \text{mm}.$$

c) Calculation with Resolving Power

As was indicated by KOMIYAMA's method in Chap. I.2.b, the resolution differs with the magnification ratio $m$. Hence, radiography was conducted while changing the magnification ratio by the use of a suitable size of line pair, lp/mm ($v$), as square-wave test object, and the magnification factor at which the test object was no longer resolved was recorded. From the lp/mm ($v$) and the magnification ratio $m$, the size of the focal spot $f$ was calculated by the formula of RAO [64]:

$$f = \frac{m}{v(m-1)}. \tag{1}$$

Formula (1) is based on the following argument.

According to MORGAN [53], the response function $F(v)$ of the focal spot can be shown to be:

$$F(v) = \frac{\sin f d_2 (d_1 + d_2)}{f d_2 (d_1 + d_2)} = \frac{\sin \theta}{\theta} \tag{2}$$

where $v$ is the spatial frequency, $f$ the focal spot size, $d_1$ the distance between focal spot and object, and $d_2$ the distance between object and film. In this case, when $\theta = \pi$ and $F(v) = 0$, the response function will become negative. This represents the limit of the resolving power, and when introduced into formula (2) becomes:

$$\frac{v f d_2}{(d_1 + d_2)} = 1 \quad \text{or} \quad \frac{v f (m-1)}{m} = 1 \tag{3}$$

where $m$ is the magnification ratio. From this we obtain:

$$f = \frac{m}{v(m-1)}. \tag{4}$$

Hence, when the loss of resolution $v$, namely $v$ (lp/mm) which becomes negative for the object, and $m$ the magnification ratio are known, the focal spot size $f$ can be

Ltd.) are used. With such a test object, measurements can be made with ease for a focal spot of 0.06 mm.

In attempting to measure a focal spot of 0.05 mm, when 10× magnification is possible, a test object with a spatial frequency of 22 lp/mm can be used; when magnification is possible only up to 5×, a test object of 25 lp/mm will be necessary; with 3× magnification, however, an object of 29.4 lp/mm will not be resolved.

Fig. 6. Relationship between focal spot size and magnification ratio expressed as curves of resolving power. The curve of resolving power decreases in accordance with the increase of magnification ratio, and the focal spot size can be read off from the ordinate.

obtained. According to COLTMAN [20], when the response function is inverted, $v$ is equal for both square-wave and sine-wave response factors, so that formula (4) holds even when a square-wave test object is used. For the actual measurement of the focal spot, the following guide lines were devised.

In measuring a focal spot of 0.05 mm, it is necessary to use a test object finer than 21 lp/mm. The size of object required to measure a certain size of focal spot and the magnification ratio indicating the limit of resolution are shown in Fig. 6, which gives the individual curves of resolution. The lp/mm ($v$) of test objects made of gold foil 0.03 mm thick and ranging from 3.85 to 20 lp/mm (Nippon Hoshasen Kenkyusho

d) Calculation with MTF Curves

The focal spot does not show a homogeneous intensity distribution, as is obtained by the pinhole image. The focal spot size is obtained from the width of its toe cut at an intensity level of 50%, but this cannot be done in case of a fine focal spot. To measure the intensity distribution of a fine focal spot, it is necessary first to obtain the modulation transfer function (MTF) (cf. Chap. I.E) of the focal spot by means of COLTMAN's formula [20].

First, the square-wave response function of the magnification radiographic system is obtained, and this is converted to the sine-wave response function by COLTMAN's formula.

The response function of the magnification radiographic system $R(v)$ is shown by the formula

$$R(v)=F\left(\frac{b}{a+b}\cdot v\right)\cdot S_{\mathrm{f}}\left(\frac{a}{a+b}\cdot v\right)$$

where $F(v)$ is the response function of the focal spot of the X-ray tube, $S_{\mathrm{f}}(v)$ is the response function of the combination of intensifying screens and film, $a$ the distance between focal spot and test object, and $b$ the distance between test object and film.

If now we change the values of $a$ and $b$, the system response functions at different magnification ratios, $R_1(v)$ and $R_2(v)$, will be indicated by the following formulae:

$$R_1(v)=F\left(\frac{b_1}{a_1+b_1}\cdot v\right)\cdot S_f\left(\frac{a_1}{a_1+b_1}\cdot v\right),$$

$$R_2(v)=F\left(\frac{b_2}{a_2+b_2}\cdot v\right)\cdot S_f\left(\frac{a_2}{a_2+b_2}\cdot v\right).$$

On elimination of $S_f$ from the above, they become:

$$\frac{R_1\left(\frac{a_2}{a_2+b_2}\cdot v\right)}{R_2\left(\frac{a_1}{a_1+b_1}\cdot v\right)}=\frac{F\left(\frac{b_1}{a_1+b_1}\cdot\frac{a_2}{a_2+b_2}\cdot v\right)}{F\left(\frac{b_2}{a_2+b_2}\cdot\frac{a_1}{a_1+b_1}\cdot v\right)}.$$

Thus, the system response functions $R_1$ and $R_2$ can be experimentally obtained and hence the exact values of $a_1$, $b_1$, $a_2$ and $b_2$ also. $F(0)$ is graphically extrapolated, and used as a normalization factor to obtain the focus response function $F(v)$.

The intensity distribution $f(x)$ of the focal spot can be computed as follows:

$$f(x)=\frac{\int_{-\infty}^{\infty}F(v)\cdot\exp(2\pi i v x)\cdot dv}{\int_{-\infty}^{\infty}F(v)\cdot dv}.$$

When using an X-ray pinhole camera the construction of the pinhole diaphragm is a problem, and this applies also when a test object is used. The accuracy of measurement will be decided by the construction of the diaphragm; it is also necessary for the arrangement of the focal spot, test object and film to be exact. In practice, however, a computer is used to calculate the approximate value by the formula below:

$$f(x)=\frac{\sum_{n=0}^{\infty}F(n\Delta v)\cos(2\pi x n\Delta v)\Delta v}{\sum_{n=0}^{\infty}F(n\Delta v)\Delta v}.$$

By plotting the values obtained, a line spread function curve can be drawn. The length at an intensity level of 50% will be the width of the focal spot. However, when a test object is used, the value will relate only to the direction vertical to the wires constituting the object, so that for obtaining the actual shape of the focal spot, it will be better to rotate the object through 45° and record the values at each direction rotated. The drawback of this method is that it depends on the thickness of the test object and the spatial frequency. It will also be necessary for the arrangement of the focal spot, test object and film to be exact.

## 3. Output

The output of tubes for magnification radiography is in general small, especially at very high magnification. This is because the focusing cup acts as a grid reducing the outflow of the electron beam from the heating filament. Moreover, if the density of the electron beam at the focal spot is too large, it will melt the surface of the target. A good magnification radiography tube is one that has a very small focal spot, emitting little off-focus radiation, and at the same time possesses a high output to enable a sharp image to be taken on the magnification radiograph at short exposure time.

An X-ray tube for low magnification radiography has a large focal spot and a relatively high output. This type of X-ray tube is used for taking magnification radiographs of moving organs, as the blurring of the X-ray image due to the penumbra induced by the relatively large size of the focal spot is cancelled out by the blurring due to the movement of the tissues.

The capacity of the X-ray tube is usually represented by the maximum output, mAs, at a fixed kV. When the mAs is increased, the exposure dose also increases linearly. Note that the smaller the focal spot, the higher the dose emitted (Table 1), due to the greater efficiency of thermoradiation from a point source.

For magnification radiography with a low output tube, the voltage of the tube should be increased so as to increase the dose of X-rays reaching the film. However, in high-magnification radiography, especially of thick parts of the body, this remedy will give a picture with poor contrast, which complicates identification of images.

Recent methods of increasing the capacity of X-ray tubes include reducing the effective focal spot by decreasing the target angle. This can be obtained also to enlarge the volume of the anode to give high heat

Table 1. X-ray tubes available for magnification radiography

| Nominal size of focal spot[a] (mm) | Rotation (RPM) | Stated frequency[b] (Hz) | Generator used (Phase) | Power rating at 1/10 sec (KW) | Maximum permissible exposure at 100 kV | Target angle | Diameter of anode desk (mm) | Manufacturer |
|---|---|---|---|---|---|---|---|---|
| 0.05 × 0.05 | 9000 | 150 | single (smoothed) | 0.45 | 120 kVp, 3 mA, 2 sec | 18° | 74 | Toshiba |
| 0.1 × 0.1 | 9000 | 150 | 3-phase 12-pulsed | 3.5 | 30 mA, 1 sec | 10° | 100 | Toshiba |
| 0.1 × 0.1 | 9720 | 180 | 3-phase 12-pulsed | 3 | 29.8 mA, 0.1 sec | 10° | 90 | Hitachi |
| 0.1 × 0.1 | 9000 | 180 | 3-phase 12-pulsed | 4.9 | 49 mA, 0.1 sec | 12° | 100 | Shimadzu |
| 0.1 × 0.1 | 8500 | 150 | 3-phase 12-pulsed | 3 | 29 mA, 1 sec | 10° | 100 | Siemens |
| 0.08 × 0.15 | 8000 (50 Hz) 9600 (60 Hz) | 150 | 3-phase 12-pulsed | 3 | 30 mA, 0.1 sec | 10° | 90 | Philips |
| less than 0.3 | 9000 | 60 | 3-phase 12-pulsed | 14 | 100 mA, 1 sec | 7° | 100 | Machlett |
| 0.2 × 0.2 | 8500 | 150 | 3-phase 12-pulsed | 6 | 51 mA, 1 sec | 10° | 100 | Siemens |

[a] Measured size of focal spot is often different from nominal size. Difference between value max. and that min. can be more than 50%.
[b] Stated frequency [150 Hz from 50 Hz mains, 180 Hz from 60 Hz mains].

capacity or to accelerate the rotation of anode to increase the cooling efficiency.

## D. Specific Effects in Magnification Radiography

The procedure for magnification radiography differs from that for normal roentgenography, as in magnification radiography the patient lies midway between the X-ray tube and the film, or closer to the tube, and the focal spot of the tube is extremely small.

This means that in interpreting the magnification radiogram X-ray images that created practically no problems or could be neglected in the normal roentgenogram, have to be analyzed because the magnification frequently makes images that are overlooked in normal roentgenography appear exaggerated. This must be regarded as a specific effect of magnification radiography.

### 1. High Detectability

#### a) Resolution

It is possible to obtain a magnified view by the optical magnifying technique instead of by magnification radiography. However, such magnified images are essentially different from those on the magnification radiograph. Although the appearance of the two roentgenograms is similar, small structures are visible in more detail on the magnification radiograph. This difference can be evaluated by a test of resolving power.

The test is based on the fact that, when two minute objects lie close together and the space between them is not visible on the X-ray image, they are not resolved. In normal roentgenography such minute spaces are frequently not visible due to the penumbra and the use of intensifying screens, but magnification radiography makes them visible as X-ray images (Fig. 7). Thus, the structure of an object that appears homogeneous in normal roentgenography is frequently imaged as a complex structure in magnification radiography. This ability to distinguish two minute objects from each other is termed resolution.

Fig. 7A–C. Resolved image of test object. Resolution of image of metal wires arranged parallel at intervals equal to the diameter of the wire in A) Normal roentgenography. B) 5 × magnification radiograph (0.032 mm, 0.045 mm, and 0.055 mm wires are resolved). C) 10 × magnification radiograph (0.025 mm, 0.032 mm and others are resolved)

Resolution is useful not only to reveal small lesions (Fig. 8), not recognizable on a normal roentgenogram, or small gaps between two tissues or lesions, but also to supply information concerning small, uneven contours of objects (Fig. 9).

Resolution is applied to show minute subjects of various sizes in high contrast. An object of 0.08 mm in high contrast is about

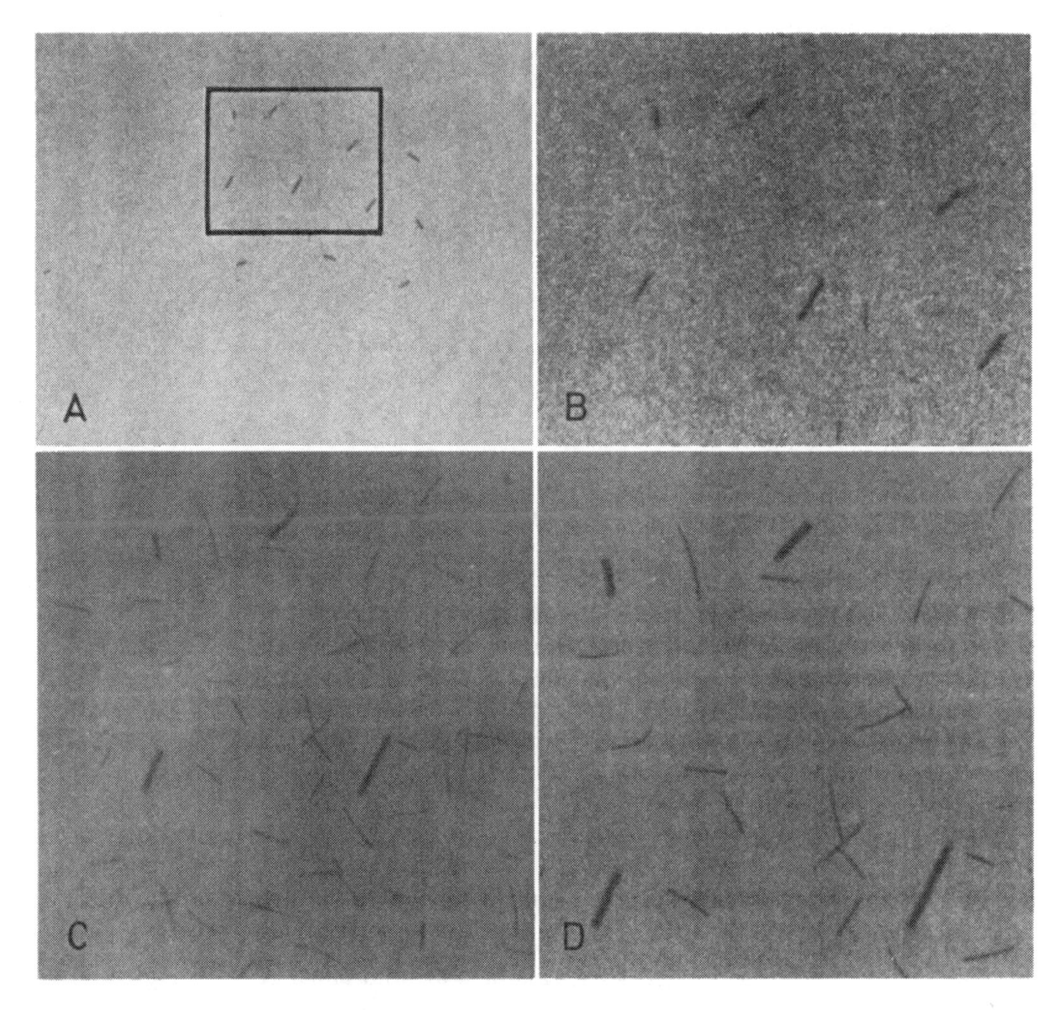


Fig. 8A–D. Improvement of detectability. Acrylic resin plate 1 cm thick on which fragments of wires of 0.049 mm in diameter which simulate capillaren or alveoli pulmonis and 0.17 mm in diameter which simulate bronchioli respiratorii or vasa lymphatica are placed. A) Normal roentgenogram taken with 2 mm focal-spot tube, 60 kV, MS screens, focal spot-film distance (FFD) 200 cm and object-film distance 15 cm. Wire fragments 0.17 mm in diameter are imaged; those 0.049 mm in diameter are not. B) 3 × optically magnified photograph of rectangular area of Fig. 8A. Fragments of wire 0.049 mm in diameter are still not imaged. C) 4 × magnification radiograph taken with 0.1 mm focal-spot tube, 120 kV, MS screens and FFD 100 cm. Fragments of 0.049 mm are visible as well as those of 0.17 mm. D) 6 × magnification radiograph taken with 0.05 mm focal-spot tube, 120 kV, MS screens and FFD 100 cm. Small fragments of wire are more easily recognized than in Fig. 8C

the limit of resolution with the naked eye in normal roentgenography. The image of an object 0.025 mm in size, even magnified 4 times, will only be about 0.1 mm in size when the background of the image is homogeneous. The resolution is thus of very small sized objects, and it is necessary to keep in mind when interpreting very small images in magnification radiography, even of high magnification, that its superiority over normal roentgenography is for objects of this grade in size.

Resolution usually applies to objects in the plane perpendicular to the central X-ray, but resolution in the longitudinal direction also plays an important role when a thick

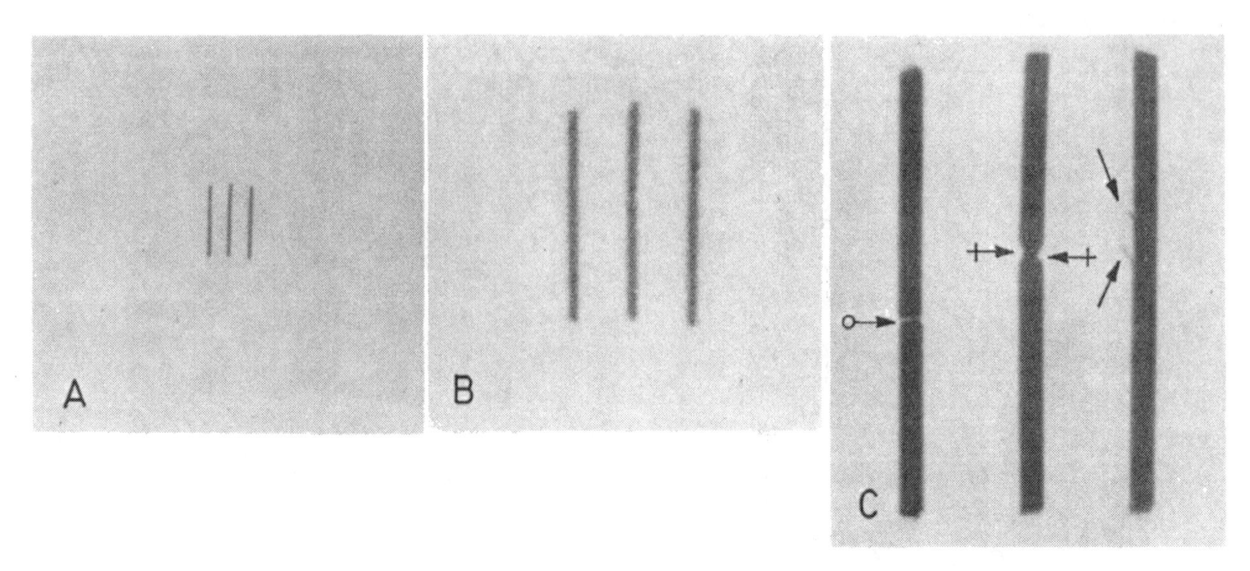


Fig. 9 A–C. Improved visualization of condition of small objects. Three wires of 0.5 mm in diameter which simulate bronchioli terminales or arterioles. *Left:* gap in wire; *center:* wire with narrowings; *right:* wire with small branches. A) Normal roentgenogram taken with 60 kV and MS screens reveals no difference in the three wires. B) 3 × optically magnified photograph of wires. Result is similar to A. C) 6 × magnification radiograph taken with 120 kV and MS screens clearly reveals gap (o→), narrowings (+→), and branches (↗) in the three wires

part of the body is subjected to magnification radiography (Doi [25]).

b) Background Dispersion

Magnification radiography is, however, not limited to the examination of minute objects. Even objects of a size that is within the limit of vision can be overlooked due to difficulty of observation caused by physiologic or psychologic factors. According to Gilardoni [31], the limit of vision is 0.15 mm at a viewing distance of 50 cm. For comfortable viewing, the smallest detail should be twice this value. Hence, details magnified to 0.3 mm or more will become easy to see. This is a facilitating of observation by magnification rather than by resolution.

According to Spiegler [94], Schwammzusatz (addition of a sponge plate) to the homogeneous background makes it difficult to recognize the image of an object in low contrast. Takahashi [119] has termed this background effect. The background effect means that the image of a small object is recognized with great difficulty when it overlaps with the image of an irregular network of small mesh. Magnification radiography resolves this phenomenon because it enables the object to be magnified in higher contrast.

This procedure is termed background dispersion. Background dispersion is useful in cases where fairly small objects are located in tissues of complex structure, since their detection by normal roentgenography is often hindered by psychological or mental factors (Fig. 10). This psychological effect can be overcome by facilitating observation by magnification. In normal roentgenography when lesions of 0.3 mm diameter are present in isolation against a homogeneous shadow, the X-ray image is easily inspected,

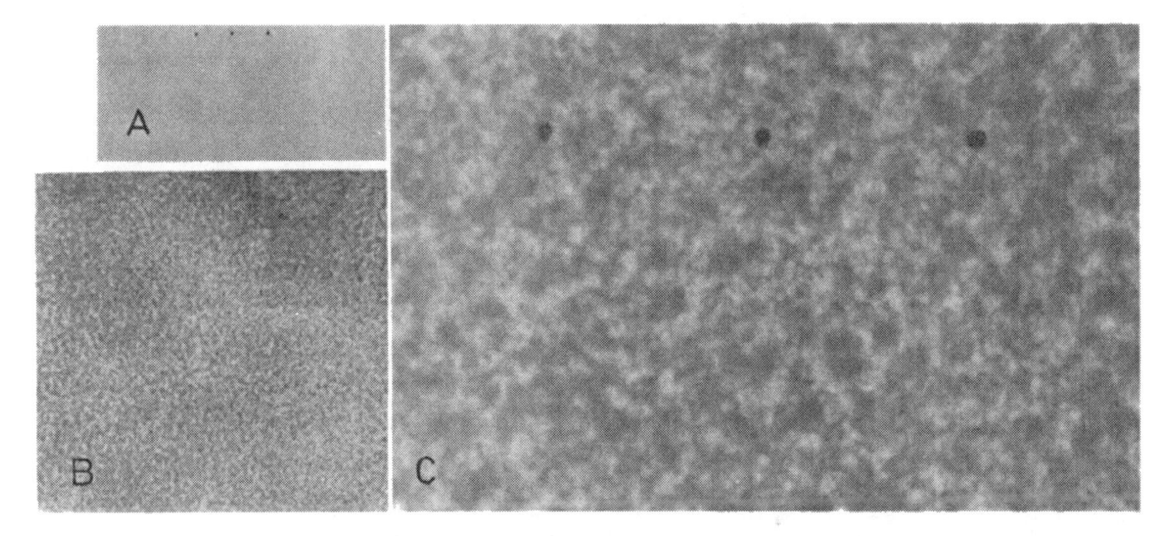


Fig. 10A–C. Background dispersion. A) Three metal beads 0.3 mm in diameter are placed on an acrylic resin plate 2 cm thick. On the normal roentgenogram of the plate the three beads are clearly imaged and visualized. B) The acrylic resin plate is placed on a sponge plate 1 cm thick soaked in barium sulfate solution. The normal roentgenogram reveals disseminated point shadows. The image of the three beads is visualized and differentiated only with the utmost difficulty. C) 6× magnification radiograph. The structure of barium grains in the sponge plate against which three beads are imaged is clearly visualized

but against a complex background it tends to be overlooked. When 0.3 mm lesions are magnified 6× to images of 1.8 mm, they become easily observable even against a background of complex structure. Background dispersion in magnification radiography can also be applied to detect a small, though not ultra-small lesion, or a small gap between two tissues or lesions, and also to obtain information concerning small, uneven contours of objects (Fig. 11). However, a too much magnified view will make it difficult to observe or detect a lesion due to poor acuity. The image of a lesion of area 1 $cm^2$ is most easily visible when the distance between the eye and the image is 60 cm. It would be absurd to subject objects large enough to be seen by normal roentgenography to magnification radiography.

Thus, improved resolution and background dispersion are considered the merits of magnification radiography.

## 2. Disappearance of Image of Very Small Object

When a minute lesion is present but is not imaged on a magnification radiograph, the phenomenon is analogous to that of small grains of sand being filtered through a wire mesh, leaving only coarse objects on the surface of the mesh. This phenomenon is thus termed filter effect [109] and it happens when the magnification is beyond the capacity of resolution. Filter effect is seen routinely in normal roentgenography and is considered mainly due to the intensifying screens. For instance, when the tip of an extremity is roentgenographed with a particular combination of film and intensifying screen, the fine part of the bone trabeculation is not imaged as well as without an intensifying screen due to the filter effect (see Figs. 40 and 41). In magnification radiography, however, the filter effect of screens does not play a significant role.

a) Filter Effect Due to Penumbra

The focal spot is not a point but occupies a certain space, therefore penumbra occurs in the X-ray image.

As discussed in C.2.b, an object of size less than one half of the focal spot will be resolved on a radiograph of sufficiently high magnification, although there may be some lack of sharpness (see Fig. 5). According to Doi [25] if the object is the same size as or smaller than the focal spot, image contrast is improved on magnification radiographs at higher magnification and helps recognition of the image. An object which is less than one half the size of the focal spot becomes faint or is filtered off from the magnification radiograph. Objects larger than the focal spot will appear in the magnification radiograph.

Clinically it is often the case that findings look similar on magnification radiographs, for instance, of a bone taken with 0.3 mm and with 0.1 mm focal-spot tube but actually when they are compared with each other on one film viewer, there is seen to be a marked difference in fine trabeculations of the bone (see Fig. 42 B and C).

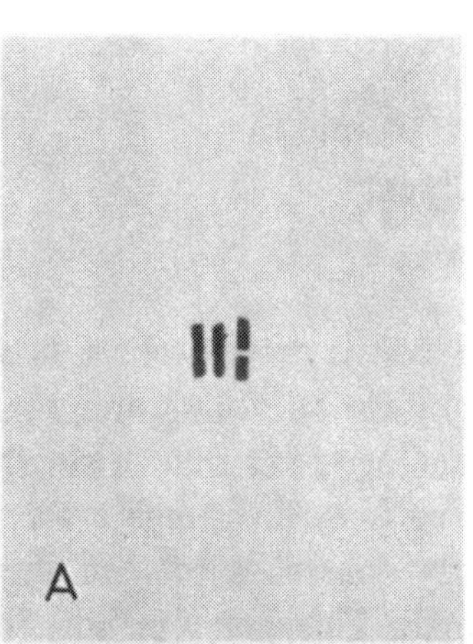


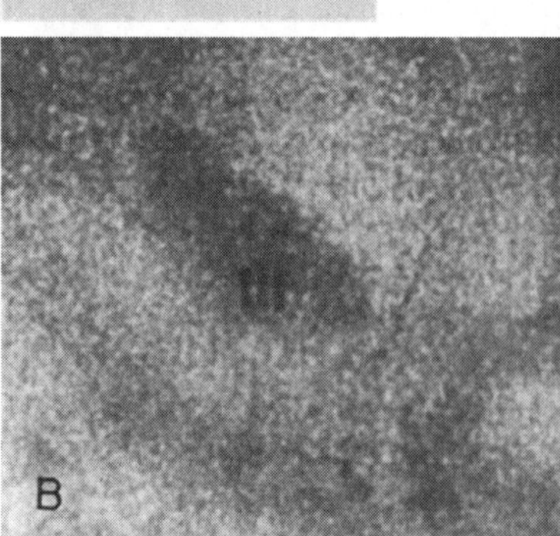


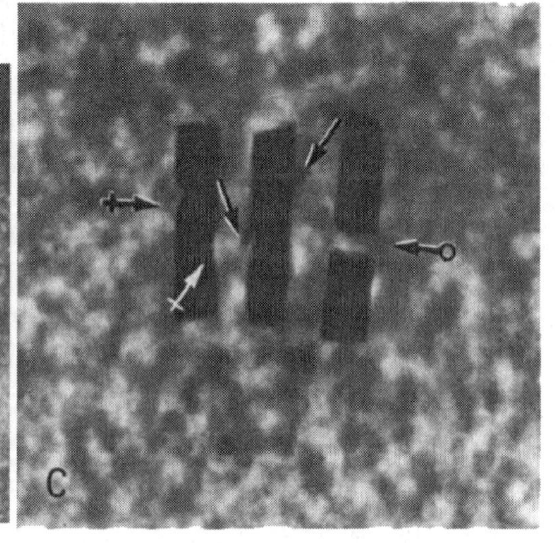


Fig. 11 A–C. Background dispersion phenomenon applied to small objects. Three wires of 1.2 mm in diameter which simulate bronchioli, small bronchus or small arteries: *Right:* wire with gap; *center*: wire with two small branches; *left:* wire with narrowings. A) Normal roentgenogram. B) The wires are placed on a sponge plate 2 mm thick soaked in barium sulfate. Normal roentgenogram taken with 60 kV and MS screens reveals only that the three wires are similar in shape. There are no other findings due to the background network. C) 6 × magnification radiograph of the three wires taken with 120 kV and MS screens clearly reveals the gap (↙o), narrowings (↙×), and two branches (↙)

### b) Filter Effect Due to Scattering

As a high-voltage technique is used in magnification radiography, scattering is a problem. The radiation field is usually small, especially in high magnification, so that scattering is less than when the radiation field is large. Moreover, in magnification radiography the film is at least 30 cm away from the human body, so that due to Groedel effect or the air-gap effect the scattering produces relatively little fog on the roentgenogram and makes the use of a grid unnecessary. However, scattering produces effect on the film.

Hence, in high-magnification radiography of thick objects, although the contrast of the image appears to be improved by the air gap between the body and film, the sharpness of the image deteriorates. A test object that is resolved for a definite spatial frequency when placed alone is not resolved when it is placed within the scattering medium (cf. Fig. 21). This is believed to be the cause of the deterioration of resolving power due to scattered rays in high magnification radiography of thick parts of the body.

Clinically, magnification angiography of thin parts of the body, such as hand or arm, provides an image of the fine structure of the vessels, but the fine vessels in the thick parts of the body, such as those of the pelvis or head, are coarse in the image. However, such findings are still better than those of the normal roentgenogram.

### c) Filter Effect Due to Contrast

Soft tissues are usually imaged in poor contrast, but bone or contrast medium provides an image in fairly high contrast. Nevertheless, even bone or contrast medium give poor-contrast images when thin.

In order to determine the relationship between the absorption and the thickness of a plate of copper, aluminium or acrylic resin, some experiments were carried out with X-rays of high and low voltage (TAKAHASHI and YOSHIDA [109, 111]) (Fig. 12). When a scatterer is not used, for a given value of contrast between ground density and density of object, contrast decreases with thickness in the order copper, aluminium, acrylic resin. It will be noted that copper is imaged in high contrast even when very thin, but acrylic resin is not contrasted well, even when thick. The very-low-voltage technique can be effective but is not practical. Aluminium falls between copper and acrylic resin in contrast imaging. A plate of acrylic resin 0.5 mm thick gives a density of 0.01 when the tube voltage is 60 kV, while that of aluminium is 0.1.

Contrast of 0.01 is the limit of visual recognition; when the background is not homogeneous, the limit about 0.1. Therefore an object of a material that is poor in X-ray absorption is not detected even when it is fairly thick. A tube voltage higher than 60 kV gives a much lower density. The resolving power of test objects made of copper and aluminium wire and acrylic thread was tested at gradually higher magnification ratios (Fig. 13). The resolving power of copper was found to increase with increase in magnification; that of aluminium increased up to a magnification of 4 × but not beyond, and that of acrylic thread only up to 2 × magnification. The test objects, with exception of copper, provide images in low contrast so that differentiation of blur due to penumbra from the image in low contrast becomes difficult.

These test objects simulate tapering parts that become very narrow and thin. The tapered part of an object with sufficient opacity to X-rays becomes so thin it can absorb very little X-radiation as a result the

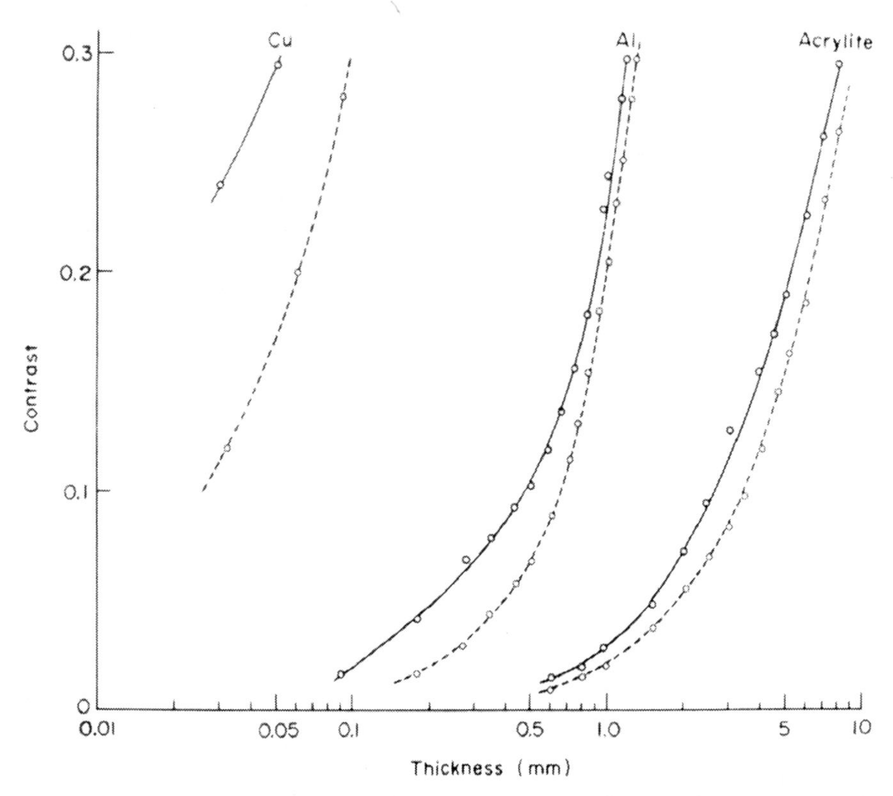


Fig. 12. Reduction in contrast of roentgenographic image of test object of copper, aluminium or acrylic resin with decreasing object thickness. ——— 60 kV, ------- 120 kV

contrast of the X-ray image is too low to be detected. This phenomenon is called the taper effect. It is a big problem in magnification radiography, the purpose of which is to image the very small and thin parts of the body.

Acrylic resin is considered to have an absorption similar to that of soft tissues, while aluminium simulates bone. Copper is equivalent to extracorporeal material such as contrast medium.

Normal roentgenograms show only bones more than 0.1 mm in diameter and soft tissues more than 0.5 mm in extent. Tapered parts fall off with decrease in X-ray absorption, so that even when magnification radiography is carried out the density of X-ray image so poor as to be undetectable. In practice, this is the limit of image formation in magnification radiography; even if magnification is attempted with a tube of minute focus it will not always be possible to visualize very minute tissues, for this is where the filter effect comes into play.

The taper effect can be avoided by increasing the contrast of the object. The low-voltage technique may contribute to obtaining an image of good contrast but this technique is not recommended to be applied with a 0.05 mm focal-spot tube as the exposure time becomes too long. In addition, the low-voltage technique is not applicable to thick parts of the body because of the high dose. The use of contrast media is better in high-magnification radiography. Images of quite small and tapered parts of tissues can be visualized by means of the high contrast obtained.

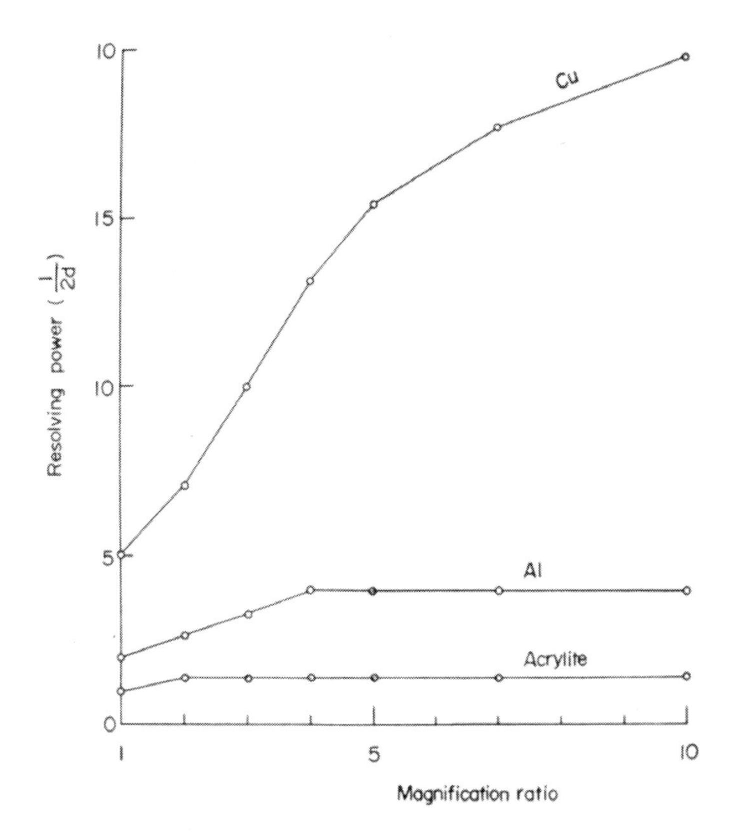


Fig. 13. Increase of resolving power in magnification radiography of test objects of copper and aluminium wire and acrylic resin thread. High voltage of 120 kV was used. In the acrylic test object, magnification radiography is not effective as regards resolving power

## 3. Magnified Blur Due to Motion

Movement of the object reduces the sharpness of the X-ray image. If the movement is substantial, the lesion can be blurred out. If the object moves during radiography at a speed of $v$ in $t$ seconds, the X-ray image will be blurred by $vt$ in normal roentgenography (Fig. 14); when it is radiographed at $\alpha \times$ magnification, the blurring will be $\alpha v t$. The lack of sharpness of the image is more noticeable in magnification radiography than in normal roentgenography. In normal roentgenography, however, the ratio of the size of blur due to motion and penumbra to the size of the image of the object is the same as in magnification radiography, so the finding itself will be equal. It is better to reduce the lack of sharpness to such an extent that the blur will not be recognized. A minute tissue or lesion is imaged originally in low

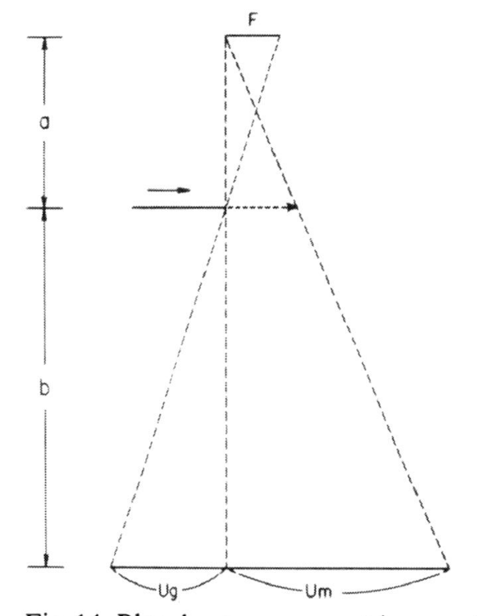


Fig. 14. Blur due to movement in magnification radiography. When the focal spot of the tube is of size $F$, the X-ray image of an object moving in the direction of the arrow will be blurred by $Ug$ (blur due to penumbra) + $Um$ (blur due to movement). $Ug$ and $Um$ are magnified $\frac{a+b}{a}$ times in magnification radiography

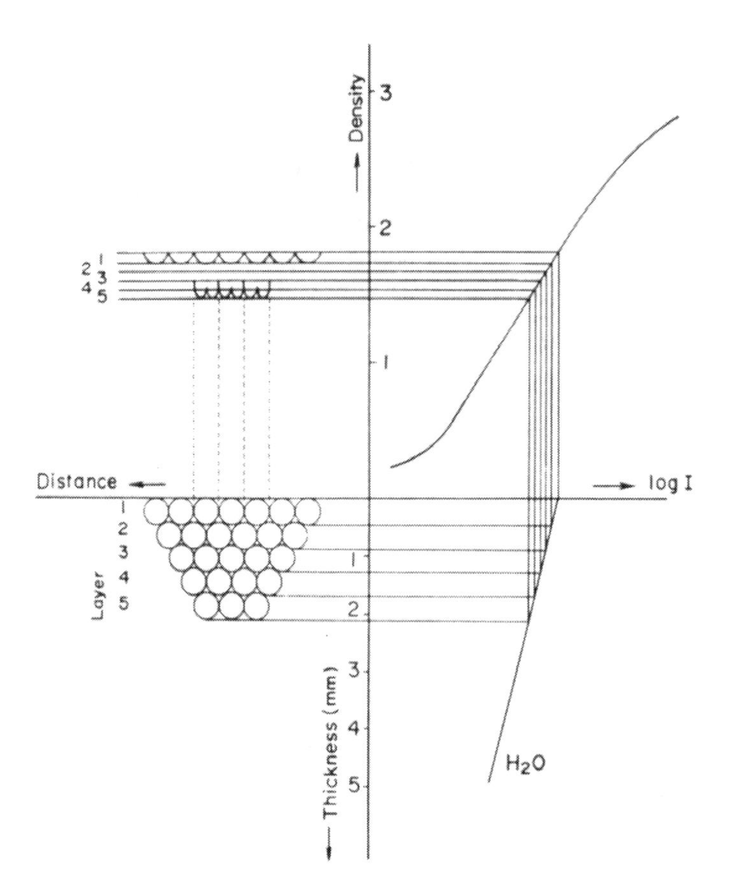


Fig. 15. Mechanism of image formation in superposition. In high magnification radiography the shape of the object becomes distorted but the contrast is improved. In roentgenographic procedure piled-up Mix D beads (bottom left) attenuate X-rays in accordance with attenuation curve of soft tissue (water) (bottom right). The attenuated X-rays blacken the X-ray film in accordance with the characteristic curve of the film (top right), so producing the image (top left). An increase in the number of beads piled up causes a higher-density image but also distortion in shape

contrast due to the taper effect. Lack of sharpness of an object in low contrast will make it psychologically difficult to recognize the object, especially in magnification. The blur should therefore be kept as small as the image in normal roentgenography. Hence, blur due to motion should be reduced to $1/\alpha$. For this, the exposure time should be $1/\alpha$. The output of the X-ray tube used for magnification radiography is anyway rather poor compared with that used in normal roentgenography, so that the reduction of mAs to $1/\alpha$ is a problem in magnification radiography.

With the exception of bone, the tissues of the body are always more or less in motion due to pulsation or peristalsis. For instance, the lower part of the left lung moves synchronously with the pulsations of the left ventricle, and this reduces the sharpness of the X-ray image. Blurring due to motion should not be greater than that caused by the penumbra or by screens, but the limiting factor for reduction of the exposure time is the output of the tube. If the body is too thick or the movement too fast, one should either use an X-ray tube of appropriate capacity or adopt a suitably low magnification ratio.

A relatively high-output X-ray tube, such as a 0.1 mm or larger focal-spot tube, will make a more effective contribution to obtaining a better X-ray image than the use of a 0.05 mm focal-spot tube. For celiac or cerebral angiography, a 0.1 mm focal-spot tube gives good results. The rotating-anode tube used for magnification radiography rotates at speeds of 3000 to 9000 rev/min or more. This rotation induces vibration, which affects the sharpness of the image. The vibration should be reduced as much as possible.

## 4. Improvement of Visibility Due to Superposition

The superposition of X-ray images sometimes makes it difficult to establish a correct diagnosis by normal roentgenography. In magnification radiography, although the circumstances are the same as in normal roentgenography, this phenomenon may help the interpretation of the magnification radiograph because of the increased contrast and the improvement in resolution and background dispersion.

Very small objects or tapered parts of tissues are not imaged, even on a magnification radiograph, due to the taper effect. They can, however, be visualized when they are piled up and the X-ray images of very small objects are superposed. The superposed image looks homogeneous on the normal roentgenogram, but on the magnification radiograph numerous irregular shadows of very small objects, different in shape from the original, are visualized. This is called the "superposition effect".

Fig. 15 illustrates the mechanism of imaging piled-up beads. The circles at lower left represent round objects piled up in layers, each layer of objects being fitted between and above the contiguous objects of the layer below. The attenuation of X-rays due to summation of thickness of the piled-up objects is shown at lower right. After the X-ray has penetrated the objects, its intensity becomes $\log I$ on the abscissa. A characteristic curve of the X-ray film is shown at upper right, with the intensity of the X-rays plotted above it. The mechanism for image production on the radiograph is shown at upper left.

When the objects are small and few, the image is in poor contrast, but as the number of small objects is increased the X-ray image of superposed objects shows medium con-

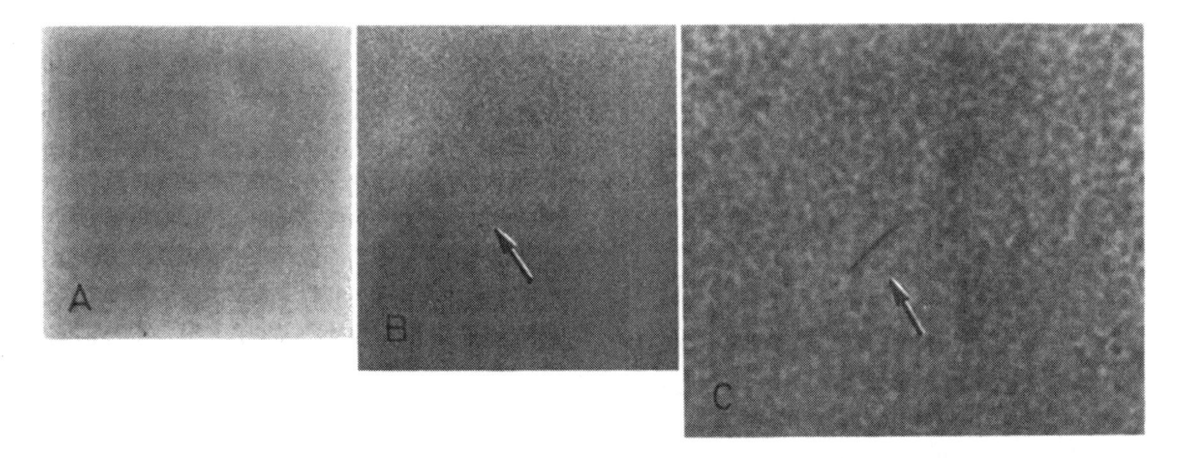


Fig. 16A–C. Improved detectability due to superposition. Salt crystals (NaCl) 0.021 mm to 0.07 mm in size are piled up to a total thickness of 3 mm. A thin wire laid on top. A) Normal roentgenogram: only the coarse, homogeneous image is visualized. The wire is not imaged. B) 2× magnification radiography conducted with 0.3 mm focal-spot tube: structure of image, though still inadequate, is better visualized than in A. Wire is faintly visualized (↗). C) 6× magnification radiograph: grains are clearly visualized, although their original shape and size are not reproduced by these images. The single wire is clearly seen in the center (↗)

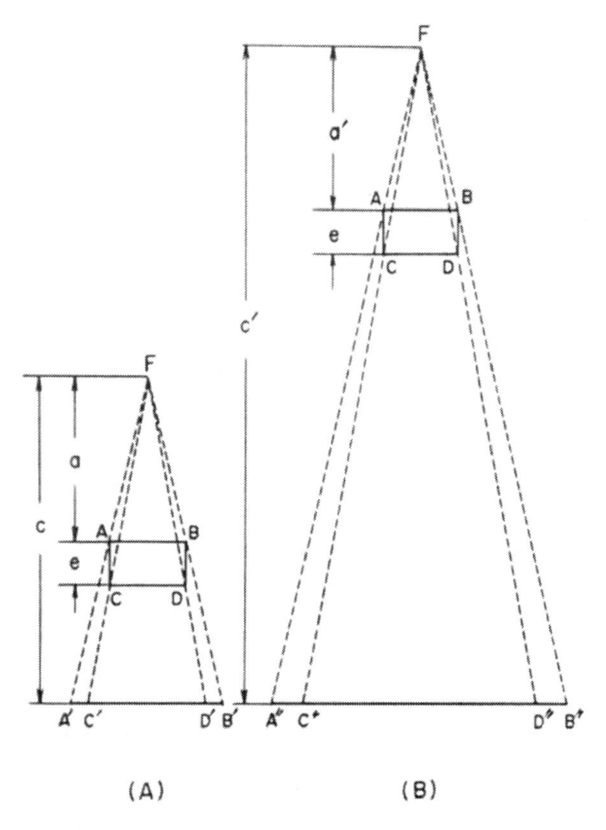


Fig. 17A and B. Variability of magnification ratio with changing distances between focal spot, object, and film. Magnification radiography with focal spot (*F*), object (*ABCD*), and film. A) Difference in the magnification ratio of $A'B'$ and $C'D'$:

$$\vartheta = 1 + \frac{e}{a}.$$

B) Difference $\vartheta'$ in magnification ratio of $A''B''$ and $C''D''$:

$$\vartheta' = 1 + \frac{e}{a'}$$

trast. It is clear from this diagram that the X-ray image does not become homogeneous, but the X-ray image is distorted as compared with the original single object.

The superposition effect in magnification radiography differs from that in normal roentgenography. The only problem here is the distortion of shape, which should be allowed for in interpretation. The magnification radiograph shows the X-ray image of a small object and can contribute to the diagnosis (Fig. 16).

## 5. Variability of Magnification Ratio

Normal roentgenography is conducted with the longest possible focal spot-film distance to keep the change of magnification ratio between various tissues in the body as small as possible. In magnification radiography there is usually some difference in the magnification ratio of body tissues near and far from the tube. If the focal spot-film distance is lengthened, so keeping the magnification ratio small, that difference becomes much less. The difference is, however, exaggerated with increasing magnification.

The focal spot-film distance is usually short in magnification radiography, so that the difference in the magnification ratio of near and distant tissues is greater than in normal roentgenography. This fact cannot be neglected in clinical practice.

Let us consider the variability of the magnification ratio. In Fig. 17A two objects, *AB* and *CD*, arranged parallel to each other at distance $e$ are imaged on the film as $A'B'$ and $C'D'$. The distance between focal spot $F$ and $AB$ is $a$ and that between $AB$ and the film is $c$. The magnification ratio of $AB$ to $A'B'$ is $\alpha_1$ and that of $CD$ to $C'D'$ is $\alpha_2$. Then

$$\alpha_1 = \frac{c}{a}$$

$$\alpha_2 = \frac{c}{a+e}$$

The difference in the magnification ratio of two objects at *AB* and *CD* is thus

$$\vartheta = \frac{\alpha_1}{\alpha_2} = \frac{\frac{c}{a}}{\frac{c}{a+e}} = \frac{a+e}{a} = 1 + \frac{e}{a}$$

From the above formula it is evident that with a fixed distance between the focal spot and the film, the shorter the focal spot-tissue distance and longer the distance between two tissues, the greater will be the difference in the magnification ratio of the two tissues.

When the distance $e$ is fixed and the distances $a$ and $c$ are changed to $a'$ and $c'$, the values $\alpha_1$ and $\alpha_2$ of the magnification ratio will change to $\alpha'_1$ and $\alpha'_2$ as follows (Fig. 17B):

$$\alpha'_1 = \frac{A''B''}{AB} = \frac{c'}{a'}$$

$$\alpha'_2 = \frac{C''D''}{CD} = \frac{c'}{a'+e}$$

From these we get the difference $\vartheta'$ in the magnification ratio of the two objects:

$$\vartheta' = \frac{\alpha'_1}{\alpha'_2} = \frac{\frac{c'}{a'}}{\frac{c'}{a'+e}} = \frac{a'+e}{a'} = 1 + \frac{e}{a'}$$

Thus,

if $a' = a$, then $\vartheta' = \vartheta$
if $a' > a$, then $\vartheta' < \vartheta$ and
if $a' < a$, then $\vartheta' > \vartheta$.

It is now evident that when the distance between two tissues in the body is fixed and the distances focal spot-film and focal spot-object are changed, the value of the difference in the magnification ratio of the two tissues will change in accordance with the focal spot-object distance (Fig. 17B).

Fig. 18 illustrates the importance of the variability of the magnification ratio in magnification radiography of thick parts of the body.

## E. Image Quality

The image quality of a roentgenogram has so far been discussed mostly in terms of sharpness and contrast. The image is, however, the product of both factors combined, and they should not be considered independently of each other.

With this in mind, a metal plate of given thickness containing line pairs, or parallel lines arranged at intervals that gradually become narrower, was used as the test object for roentgenography and magnification radiography. If the contrast $C$ between metal plate and interval is defined as the ratio between the sum and the difference of the maximum density ($D_{max}$) and the minimum density ($D_{min}$) of the image of the interval, then

$$C = \frac{D_{max} - D_{min}}{D_{max} + D_{min}}.$$

With this ratio as the $y$ axis, and the values of the gradually narrowed interval expressed as lp/mm (or spatial frequency) as the $x$ axis, a square-wave response function curve can be drawn. This can be converted to a sine-wave response function by the method of Coltman [20]. This function, called the modulation transfer function (MTF) or response function, indicates the relationship between the size of the object radiographed and the precision of the information recorded on the radiograph, in other words, the quality of the X-ray image.

When two MTF's cross at a certain point, this means that at frequencies above and below this point the amount of information

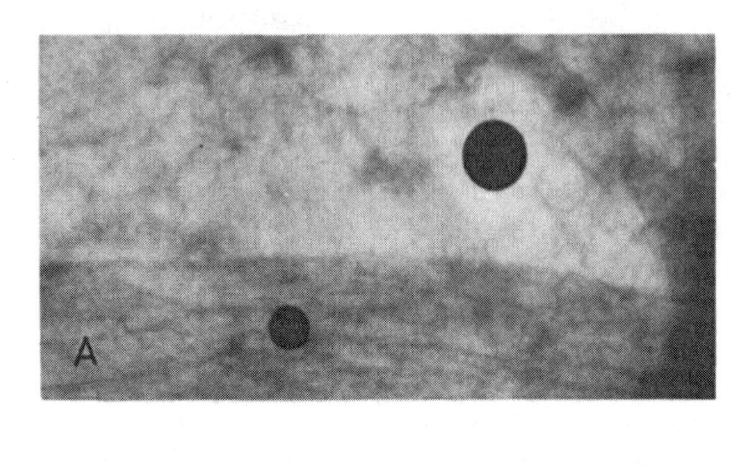


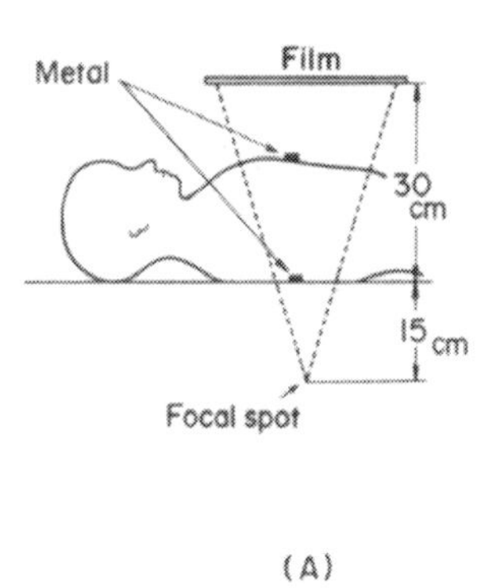


(A)

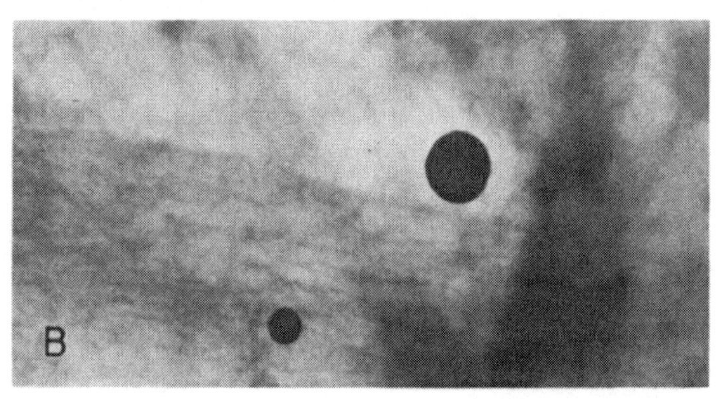


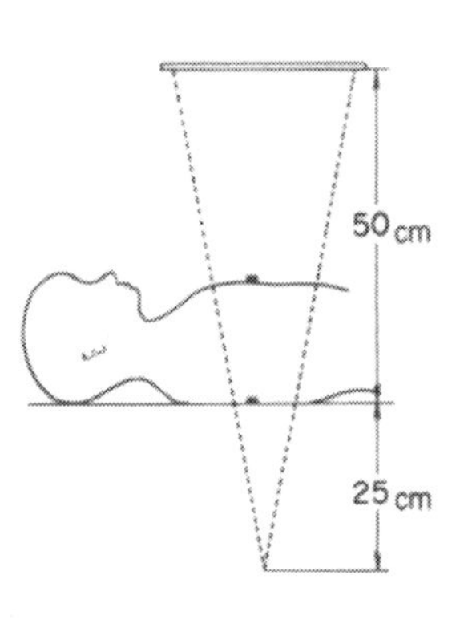


(B)

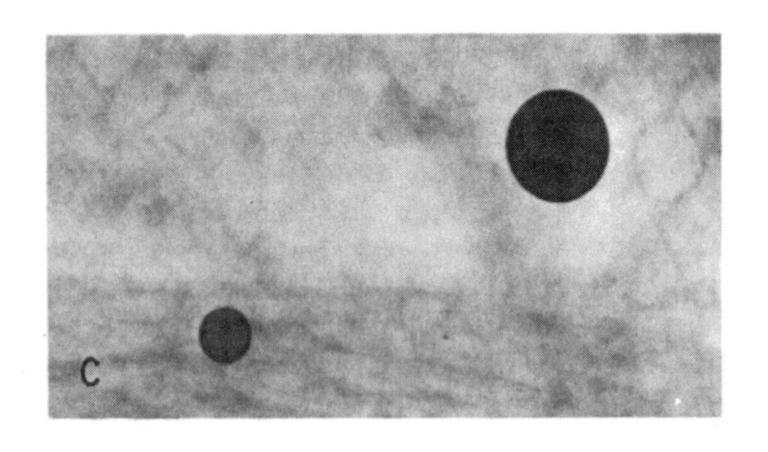


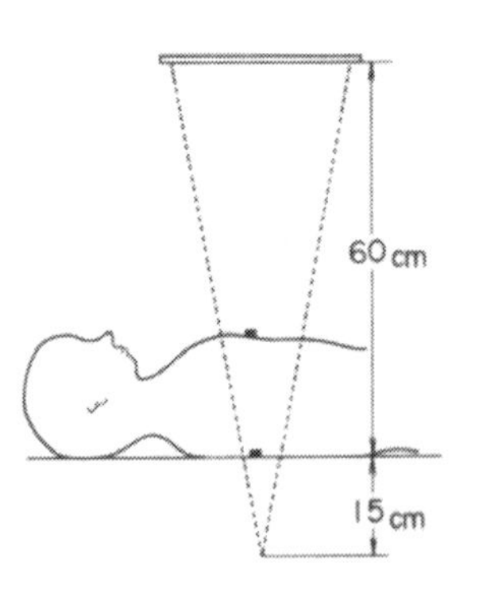


(C)

provided by the radiographic system in question will be reversed. The information capacity is proportional to the area circumscribed by the curve of the MTF and the $x$ axis. In general, the reproduction of large objects will be good in the low-frequency area and the reproduction of fine objects will be restricted in the high-frequency area. The point where the MTF curve crosses the $x$ axis is believed to represent the resolving power.

The quality of the image on a magnification radiograph is influenced by the size of the focal spot of the X-ray tube, the resolving power of the intensifying screens, the speed of the film, etc. This is well known and has also been clarified by many researchers studying the curves of the MTF.

However, in actual practice the problems arising in magnification radiography of the human body are influenced, in addition to the above, by the combined effects of the thickness of the part of the body to be examined and the movement of its organs.

The response function of a radiographic system is given by the formula:

$$R(v)=F\left(\frac{b}{a+b}\cdot v\right)\cdot S_{\mathrm{f}}\left(\frac{a}{a+b}\cdot v\right) \quad (1)$$

where $F(v)$ is the response function of the focus, $S_{\mathrm{f}}(v)$ the response function of the combination of intensifying screens and film, $a$ the focal spot-object distance, $b$ the object-film distance, and $v$ the spatial frequency, in this case, the size of the object to be examined in the body (SAKUMA *et al.* [78]). When the thickness of the object is taken to be $d$, its response function per unit as $D(v)$ and the response function of movement as $M(v)$, then the response function of the object, $O(v)$, will be

$$O(v)=D(v)^{\mathrm{d}}\,M(v) \quad (2)$$

and the response function of the image $I(v)$ will be:

$$I(v)=R(v)\cdot O(v). \quad (3)$$

Experiments were carried out with the following X-ray tubes: 0.05 mm focal spot (Toshiba DRX-90H) with output of 3 mAs; 0.1 mm focal spot (Hitachi UH-6H-10) with output of 20 mAs; and 0.3 mm focal spot (Toshiba DRX-191A) with output of 40 mAs. Medium-speed intensifying screens (Kyokko MS) and highly sensitive medical X-ray film (Kodak RPR) were used at the tube voltage of 120 kV.

The system response functions $R(v)$ obtained when the magnification ratio of the magnification radiography system was changed are shown in Figs. 19–23. Where no

◁
Fig. 18 *A*–C. Variability of magnification ratio in magnification radiography. *Right:* Schema of positioning of patient. Two metal beads are placed on the chest and back of a supine patient whose body is 18 cm thick in a magnification radiography unit of under-tube type (see Fig. 30). Magnification radiography is performed with distances: A) Focal spot-table 15 cm; table-film 30 cm: magnification ratio 3. B) Focal spot-table 25 cm; table-film 50 cm: magnification ratio 3. C) Focal spot-table 15 cm; table-film 60 cm: magnification ratio 5. *Left:* Radiographs taken as specified in A, B and C above 1. A and B had the same magnification ratio with a different focal spot-film distance; comparison shows that the difference in magnification ratio is less in B. 2. B and C had a different magnification ratio with the same focal spot-film distance; comparison shows that the difference in magnification ratio is much greater in C. 3. A and C had a different magnification ratio with the same focal spot-table distance. Here the difference in magnification ratio is the same.

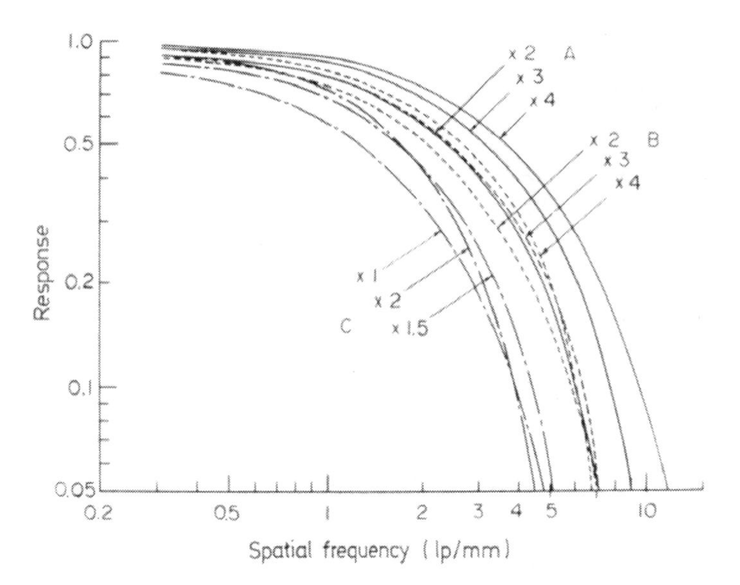


Fig. 19 A–C. Modulation transfer functions (MTF) for magnification radiography with A) 0.05 mm focal-spot tube (solid line); B) 0.1 mm focal-spot tube (dashed line); C) 0.3 mm focal-spot tube (dashed and solid line) under the exposure conditions of 120 kV, FFD 100 cm, medium-speed screens (Kyokko MS) for the test object of Optiker Funk, Erlangen, which is placed on the table without scatterer

scatterer is used, 4 × magnification is optimal with a 0.05 mm focal-spot tube, 3 × magnification with a 0.1 mm focal-spot tube and 1.5 × magnification with a 0.3 mm focal spot (Fig. 19). A comparison of the MTF curves shows that the smaller the focal spot, the greater the information capacity. These system response functions for different magnification ratios can be used to calculate the response functions of each focal spot and of the combinations of screens and film.

The MTFs of the magnification radiography system at different magnification ratios, referred to by the suffixes 1 and 2, respectively, are given by

$$R_1(v) = F\left(\frac{b_1}{a_1+b_1}\cdot v\right) \cdot S_f\left(\frac{a_1}{a_1+b_1}\cdot v\right) \tag{4}$$

and

$$R_2(v) = F\left(\frac{b_2}{a_2+b_2}\cdot v\right) \cdot S_f\left(\frac{a_2}{a_2+b_2}\cdot v\right). \tag{5}$$

Elimination of $S_f$ from these equations gives

$$\frac{R_1\left(\frac{a_2}{a_2+b_2}\cdot v\right)}{R_2\left(\frac{a_1}{a_1+b_1}\cdot v\right)} = \frac{F\left(\frac{b_1}{a_1+b_1}\cdot\frac{a_2}{a_2+b_2}\cdot v\right)}{F\left(\frac{b_2}{a_2+b_2}\cdot\frac{a_1}{a_1+b_1}\cdot v\right)}. \tag{6}$$

If two sets of tests are performed and the distances $a_1, b_1$ and $a_2, b_2$ are determined, the system response functions $R_1$ and $R_2$ at a definite spatial frequency can be obtained experimentally. Thus, the ratio on the left-hand side of Eq. (6) is obtained. Eq. (6) shows that the ratio of system response functions should be equal to that of focal-spot response functions. The ratio of system response functions appearing in Eq. (6) has been obtained as a function of $v$ by MTF's of the magnification radiography system obtained at 2 × and 4 × magnification as shown in Fig. 19. $F(0)$ was extrapolated graphically and was used as the normalization factor.

The MTF's obtained for various sizes of spot and the MTF of the combination of intensifying screen and film are shown in

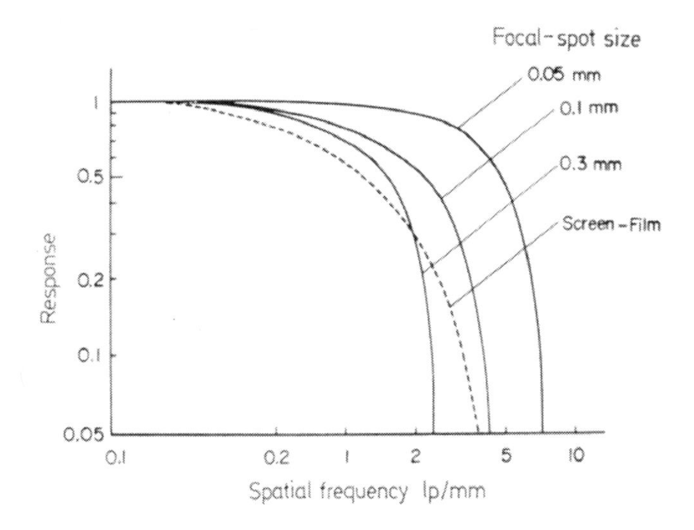


Fig. 20. Modulation transfer functions (MTF) of 0.05 mm, 0.1 mm or 0.3 mm focal-spot tubes. The dashed line represents the MTF of the combination of screens (medium-speed) and film (RPR)

Fig. 20. The solid lines show the MTF's of focal-spot sizes 0.05 mm, 0.1 mm and 0.3 mm; the dashed line is the MTF of the screen-film combination. In order to get the value of MTF in $\frac{a+b}{a}\times$ magnification radiography, the actual numerical values of response are obtained at a given spatial frequency, where $a$ is the focal spot-object distance and $b$ is the object-film distance. The values of $a$ and $b$ are inserted into Eq. (1) and calculated.

With the object placed in the scatterer, the response function $R(v) \cdot D(v)^d$ can also be calculated. The MTF curves obtained by actual experiment in radiography at various levels of magnification are shown in Fig. 21. The curves in the left column were obtained with a 0.05 mm focal-spot tube and water phantoms 5, 10, 15 and 20 cm thick at 4, 6, 8 and 10 × magnification. The MTF curves in the center and right columns were obtained under the same conditions with the 0.1 mm and 0.3 mm focal-spot tubes, respectively.

The conclusions from the experiments made with the scatterer are that the response function becomes optimal when the scatterer is inserted between the tube and film, for a 0.05 mm focal-spot tube at 6 to 8 × magnification, for a 0.1 mm focal-spot tube at 3 to 4 ×, for a 0.3 mm focal-spot tube at 2 × or slightly more.

When conducting magnification radiography of various magnification ratios with various thicknesses of water phantom, we recorded the response function of the scatterer $D(v)^d$ for water thickness at 5 cm intervals and then calculated the response function per cm water from the response function of the scatterer at 5 cm water. The thickness of the object is determined, the value is converted in terms of thickness of water, and the effect of thickness when the object is radiographed at various magnification ratios can then be calculated. Fig. 22 shows the bad effects of a scatterer on image quality per cm of water with normal roentgenography and with 1.5, 2, 3, 4, 6 and 8 × magnification radiography at FFD 100 cm. These curves enable the MTF of a scatterer for a given thickness to be simply calculated by means of Eq. (2).

It will be seen from the figures that in the case of normal roentgenography with a scatterer there is a fall in response throughout the entire spatial frequency range of the object, which becomes very marked in the high-frequency range. With increasing

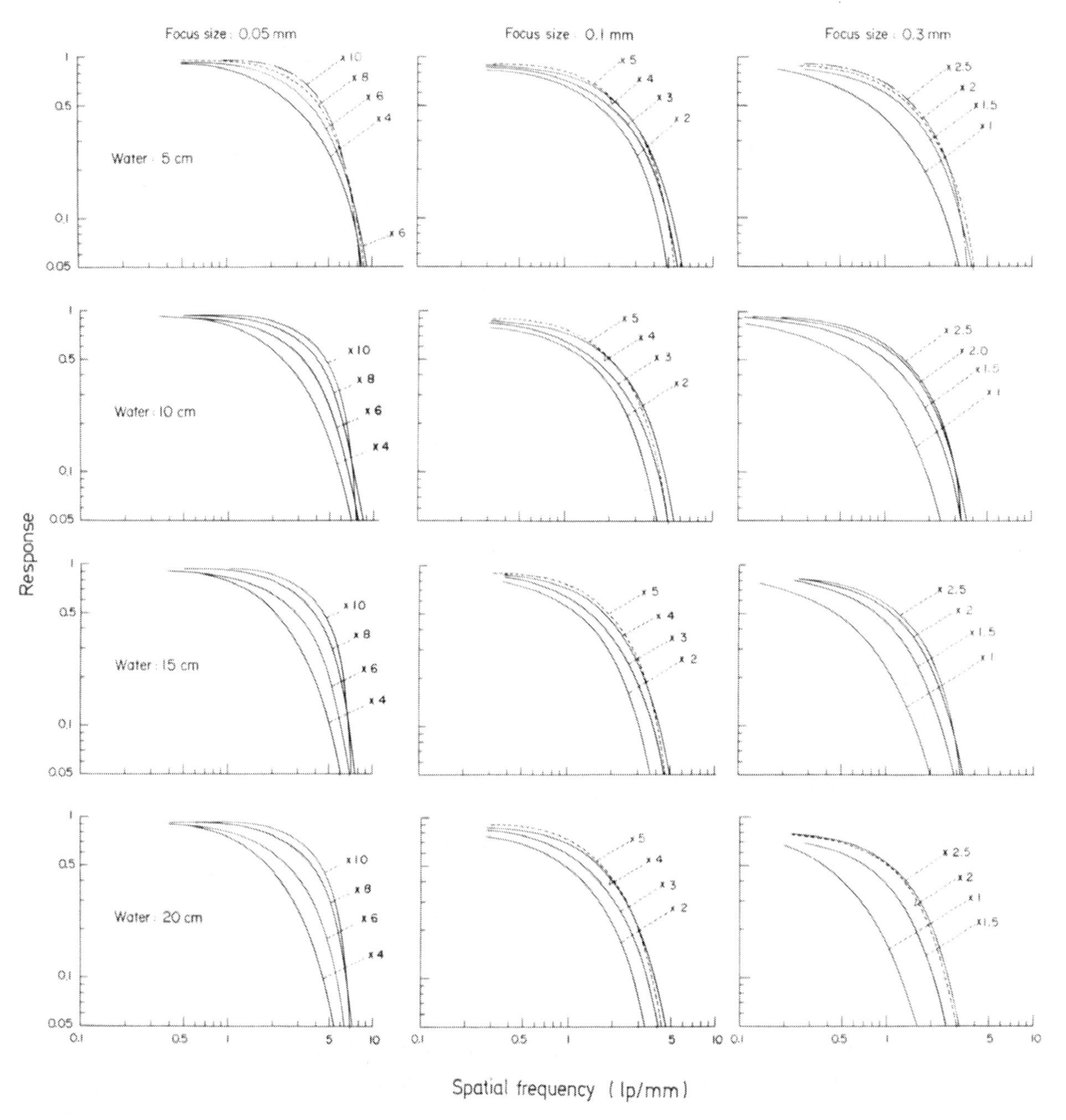


Fig. 21. Modulation transfer functions (MTF) of magnification radiographs of test object (Optiker Funk) through 5 cm, 10 cm, 15 cm or 20 cm of water, taken by X-ray tubes having 0.05 mm, 0.1 mm or 0.3 mm focal spot. For a given size of focal spot, the thicker the phantom of water, the poorer the response value. For a given thickness of water, the larger the focal spot, the poorer the response value

magnification ratio the fall in response becomes smaller, due to the effects of the air gap.

The capacity of an X-ray tube declines as the focal spot decreases in size, so that exposure time has to be prolonged for a thick object and the problem of motion effect arises. In magnification radiography of minute objects within the body it thus becomes necessary to bear these factors in

Fig. 22. Modulation transfer functions (MTF) of magnification radiographs of test object (Optiker Funk) at various magnifications through 1 cm of water. MTF curves shift toward higher spatial frequencies with increasing magnification ratio

mind when deciding on the optimal technique and radiographic conditions.

When an object moves at uniform speed, the response function of movement can be obtained (MORGAN [53]). Needless to say, the exposure time should be as short as possible, but it will also be influenced by type of examination, film, speed and capacity of the tube. When the tube voltage is 120 kVp, the maximum allowable tube current and air dose at the point on the central ray 100 cm distant from the focal spot is limited for several types of X-ray tube in magnification radiography (see Table 1). If the tube current routinely used is 80% of the maximum tube current, this will become 3 mA, 20 mA and 40 mA, respectively. The shorter the exposure time, the sharper the image of a moving object.

The response function $M(v)$ of movement is a function of the speed of movement of the organ and the exposure time. $M(v)$ was obtained for each speed and applied to the system response function $R(v) \cdot D(v)^{d}$ obtained in the previously mentioned experiment with a water phantom as scatterer; this gives the final overall response function $I(v)$ of the image. However, the actual experiment was first carried out, and the data thus obtained are shown in Fig. 23: the curves of the magnification ratios indicate the optimal response for each size of focal spot.

The results indicate that when the speed of the movement of the object is below 0.05 mm/sec, high-magnification radiography with a 0.05 mm focal spot will produce a high-quality image, even with a tube of poor capacity, irrespective of the thickness of the object. When the speed of movement is 0.5 mm/sec, objects up to 15 cm thick can be radiographed with a 0.05 mm focal spot, but with thicker objects it is better to use a tube of 0.1 mm focal spot at a lower magnification ratio. With a speed of movement of 2 mm/sec, high magnification of objects less than 10 cm thick is possible with a 0.05 mm focal spot; with objects 10 to 20 cm thick a lower magnification should be attempted with a 0.1 mm focal spot; with objects more than 20 cm thick 2× magnification with a 0.3 mm focal spot is recommended.

These experiments were carried out to show the appropriate size of focal spot and magnification ratio to obtain maximal information capacity in magnification radiography of a moving tissue located within the body. For a non-moving object, the best results were obtained with a very small focal-spot tube. An increase in speed of movement of the tissue did not necessarily require a small size of focal spot; rather,

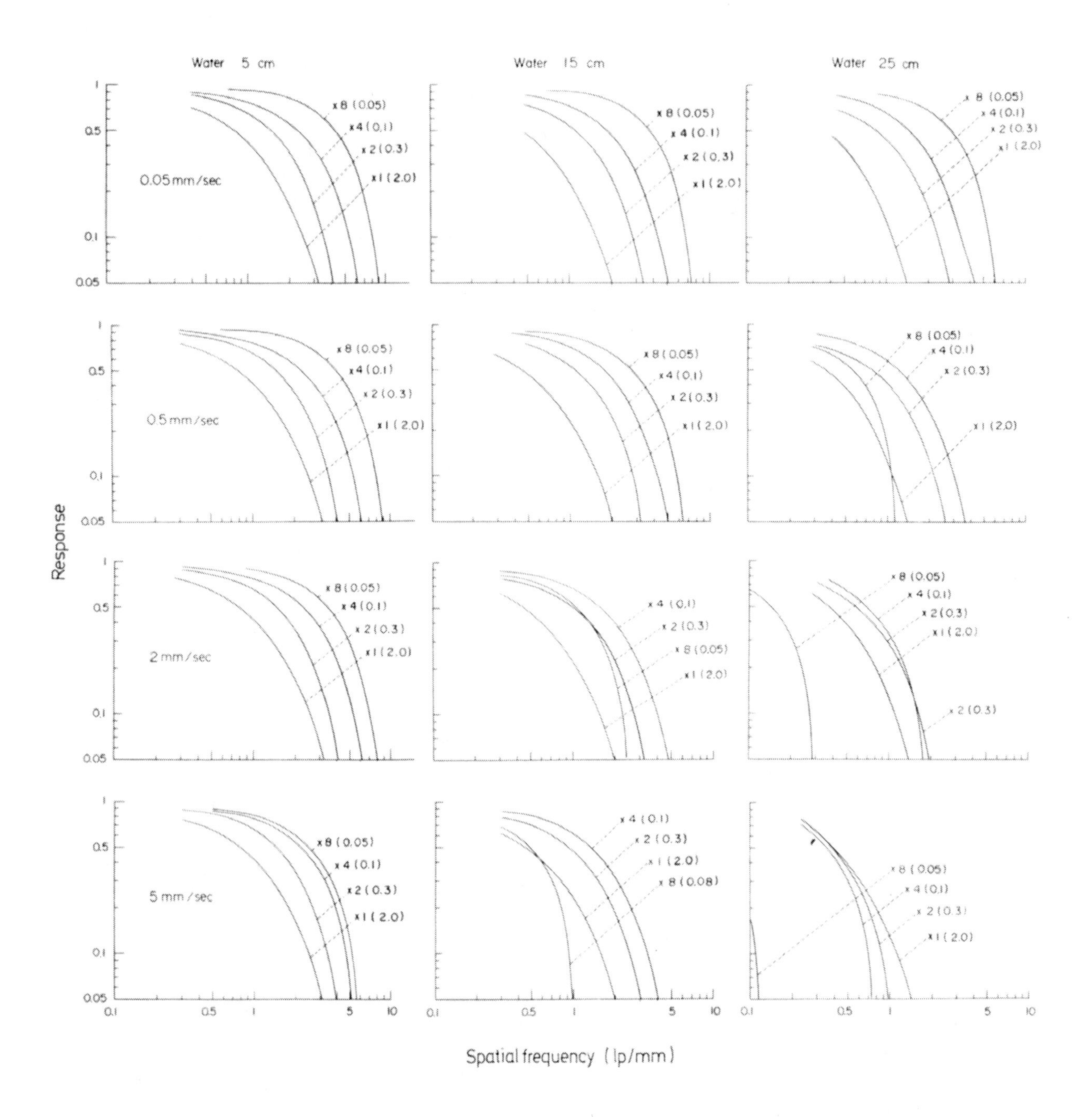


Fig. 23. Modulation transfer functions (MTF) of magnification radiographs of test object (Optiker Funk) moving at a uniform speed of 0.05 mm/sec, 0.5 mm/sec, 2 mm/sec or 5 mm/sec, in water 5 cm, 15 cm or 25 cm thick. Each set of radiographs was taken with X-ray tubes having 0.05 mm, 0.1 mm, 0.3 mm or 2.0 mm focal spot, of which tube current is fixed with 3 mA, 20 mA, 40 mA, or 200 mA respectively for radiography in this experiment. The best response value is obtained at 8 × magnification for a 0.05 mm focal-spot tube and 5 cm of water. MTF curves differ with speed of movement of test object, thickness of phantom, and size of focal spot. The less the thickness of water and the slower the movement of the object, the better the response value of high-magnification radiography with the smaller focal-spot tube. With magnification radiography of a moving object and a thick phantom, the response value rapidly becomes worse with increasing thickness of object. It is noted that here the small focal-spot tube has a poor capacity so that its use is limited to short exposure times.

the larger the capacity of the X-ray tube and the shorter the exposure time, the better were the results obtained.

When the exposure conditions are adjusted so as to limit the air dose at the skin to less than 1 R (cf. Chaps. I.F.6 and II.G), it is evident that 4 to 6 × magnification radiography of the chest or other thin body parts can be done with a focal spot-skin distance of 15 cm. With thicker objects like the abdomen, the magnification must be lowered to 3 to 4 × in order to reduce the skin dose. Although a 0.05 mm focal-spot tube provides a magnification radiograph of much better quality than a 0.1 mm focal-spot tube for 2, 4 or 8 × magnification of a non-moving tissue or organ in the thick parts of the body (see Fig. 21), there was less improvement in information capacity when magnification radiography was applied to a very slightly moving object in the thick part of the body. Four times magnification radiography provided a comparable amount of information with a tube having a 0.1 or 0.05 mm focal spot (see Fig. 23), while the magnification radiograph taken with a 0.05 mm focal spot tube was richer in information at higher magnification, e.g. 8 ×, than that taken with a 0.1 mm focal-spot tube. Therefore, in actual practice when it is necessary to limit the X-ray dose to the skin, the exposure conditions must be calculated by computer so as to obtain the maximum information from an image of detailed structure at the lowest possible skin dose.

Thus, with thin objects, such as the tip of an extremity or an air-rich organ like the lung, high-magnification radiography can be performed with a 0.05 mm focal-spot tube, as the exposure time is reasonably short. For the skeletal system and other parts that do not move, much higher magnification radiography with a 0.05 mm focal spot is appropriate. For thick objects that are moving, such as the abdomen and the digestive tract, a 0.1 mm or even a 0.3 mm focal-spot tube is recommended with 2 or 4 × magnification radiography. For a fast moving object, it may even be advisable to use normal roentgenography at very short exposure time, employing a tube with large focal spot.

To sum up, we should choose the radiographic technique which offers the best response to the spatial frequency of the object.

## F. Radiation Protection of the Patient

One of the main problems in magnification radiography is reduction of the dose to the patient, as the radiation exposure increases in proportion to the increase in magnification ratio.

Magnification radiography is usually conducted with a fixed distance of about 100 cm between tube focus and film. This distance is almost halved in 2 × magnification radiography, as the patient is positioned midway between the focal spot and film. For 2 × magnification radiograph having the same density as a normal roentgenogram, the skin will be irradiated about four times as much as in normal roentgenography. ZIMMER [140], one of the pioneers in the clinical application of magnification radiography, cautioned as early as in 1951 that the skin dose is high. ADERHOLD and SEIFERT [3] also stated that the X-ray dose rose in proportion to the magnification ratio and that it should therefore be limited to 4 to 6 × magnification.

Today, there is still a tendency for patients to receive undue radiation exposure in X-ray

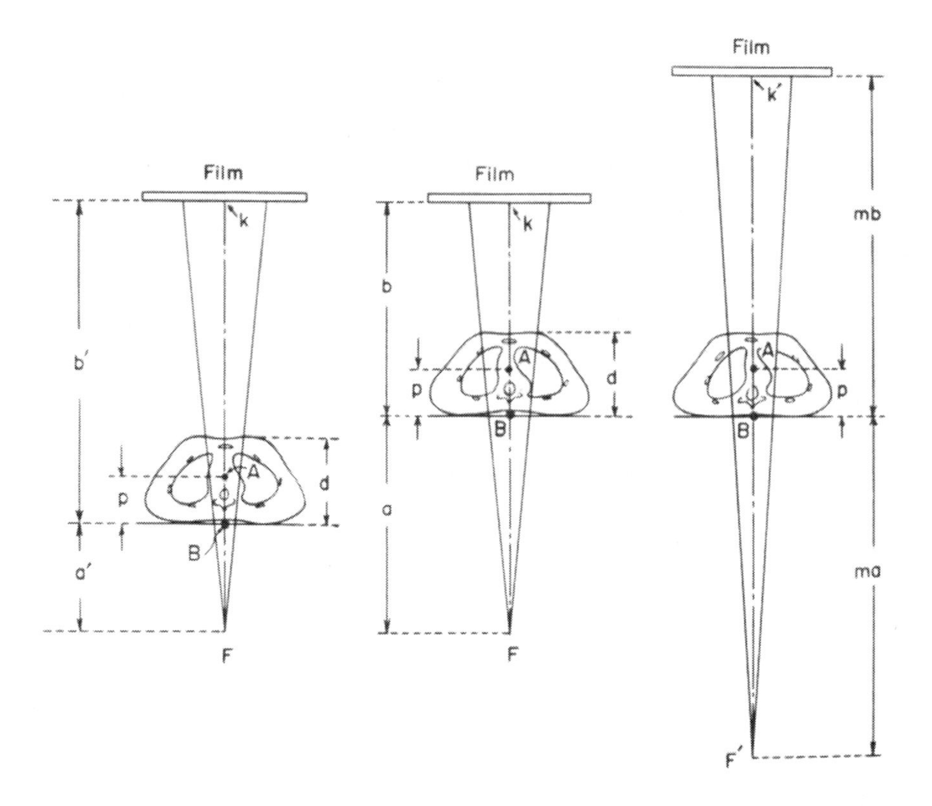


Fig. 24. Relationship between exposure dose to the skin and magnification ratio. *Center:* Skin dose is $\frac{(a+b)^2}{a^2}Q$, where $Q$ is a constant. *Left:* Skin dose is $\frac{a^2}{a'^2}$ when the focal spot-film distance $a+b$ is the same as in the center diagram. Magnification ratio is the same but skin dose is higher than in the case of the center diagram. *Right:* Skin dose is the same as in the center diagram when the focal spot-film distance and the skin-film distance are $ma$ and $mb$, respectively. Both magnification ratio and skin dose are the same as in the center diagram. *F* Focal spot; *A* Lesion; *B* Table top; *p* Lesion-skin distance; *d* Thickness of body; *k* Density of film; *a* Focal spot-skin distance; *b* Skin-film distance

examinations, and the International Commission on Radiological Protection has recommended that the patient dose be reduced to a reasonably low level. If magnification radiography is widely applied in routine clinical examinations without due caution, there is a risk of injury to the patient. The need for adequate attention to this problem cannot be overstated.

In order to avoid unnecessary exposure of the patient in magnification radiography, it is essential to have a thorough knowledge of the relationships between skin dose, tube voltage, thickness of part to be examined, depth dose, volume dose, and size of radiation field (MAEKOSHI *et al.* [49]; TAKAHASHI *et al.* [120]).

## 1. Skin Dose

If the focal-spot-film distance (FFD) is fixed at 100 cm, varying the distance between focal spot and the patient will cause a change in the magnification ratio (Fig. 24).

Actual measurement of the dose revealed that in magnification radiography, when the

FFD is kept constant, the X-ray dose at the skin surface rapidly increases at higher magnifications. In other words, the magnification ratio has to be considered from the aspect of skin dose increases.

This result is explained by the following theoretical considerations. If the distance between the focal spot and the skin facing the tube is $a$ and that between the skin and the film is $b$, the FFD is $a+b$ (Fig. 24, center). If the X-ray dose reaching the film is $K$, the X-ray dose $E$ to the patient's skin will be

$$E=K\frac{(a+b)^2\exp(\mu d)}{a^2},$$

where $\mu$ is the linear attenuation coefficient of the body and $d$ the thickness of the body.

When the distances are made $a'$ and $b'$ respectively, with $a'<a$ and $a+b=a'+b'$ (Fig. 24, left), the magnification ratio will be higher than that in the first case. The skin dose, $Ea$, will be

$$Ea=K\frac{(a'+b')^2\exp(\mu d)}{a'^2}$$

$$=K\frac{(a+b)^2\exp(\mu d)}{a'^2}$$

Hence

$$\frac{Ea}{E}=\frac{a^2}{a'^2}$$

From this it is clear that in the case of a fixed FFD the patient dose increases in accordance with the inverse-square law with increasing magnification ratio.

When the magnification ratio and the density of film obtained are kept constant, the distance between the focal spot and the patient's skin is $ma$ and that between skin surface and film $mb$, the FFD will be $m(a+b)$ (Fig. 24, right). Assuming the X-ray dose reaching the film to be $K$, the radiation exposure, $Eb$, to the patient's skin will be

$$Eb=\frac{(ma+mb)^2\exp(\mu d)}{(am)^2}K'$$

$$=\frac{m^2(a+b)^2\exp(\mu d)}{a^2m^2}K'$$

Here $K=K'$, as in order to obtain a film density of 1.0 the X-ray dose at the film surface has to have a certain value. Therefore, it is concluded that $E=Eb$. In other words, when the magnification ratio is fixed, the exposure dose to the patient's skin will be the same, irrespective of the distance between focal spot and film.

*It is thus evident that the skin dose of the patient increases as magnification ratio becomes higher in magnification radiography.*

## 2. Size of Radiation Field

In roentgenography the size of the X-ray beam should never exceed the film size, as an X-ray beam projecting beyond from the film does not contribute to image formation. In addition, it is a hazard to the patient. In magnification radiography the X-ray beam should cover the tissue of interest alone, and the size of X-ray film must be chosen for this purpose.

Since in magnification radiography the patient is positioned midway between the tube and film, the radiation field of the patient will naturally decrease as the magnification ratio is increased (Fig. 25). For a small radiation field, the exposure is somewhat higher than for a large field, but in magnification radiography the skin dose increases approximately in accordance with the inverse-square law with increasing magnification ratio.

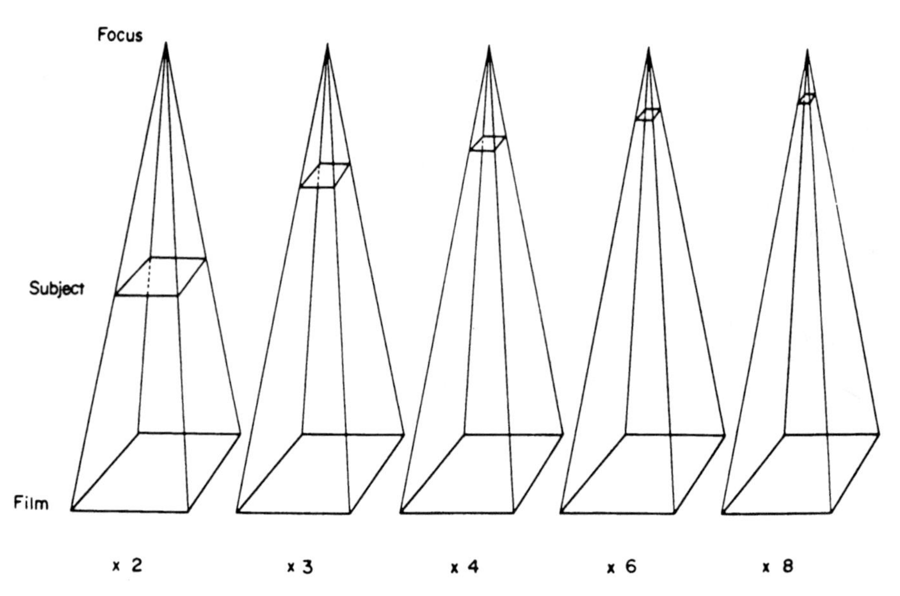


Fig. 25. Field size of magnification radiographs taken with fixed focal spot-film distance. The higher the magnification ratio, the smaller the field size in accordance with inverse-square law

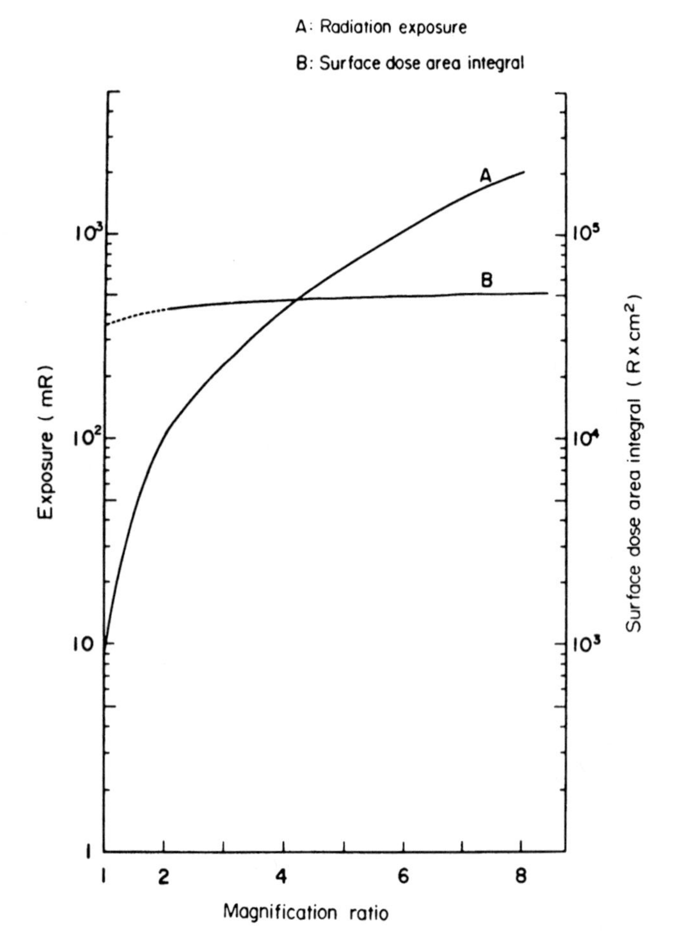


Fig. 26. Relationship between magnification ratio and exposure dose to the skin in magnification radiography. With a fixed FFD of 100 cm, the magnification ratio rises as the FSD is reduced. Note that the skin dose increases exponentially with higher magnification ratios, while the surface-dose area integral rad $\cdot$ cm$^2$ or R $\cdot$ cm$^2$ is nearly constant. A radiation field $5 \times 5$ cm at $8 \times$ magnification, $10 \times 10$ cm at $4 \times$ magnification, and $30 \times 30$ cm in normal roentgenography conducted with a grid of grid ratio 12. A water phantom of 15 cm thickness was used. Exposure factors: 118 kV with film density of 1.0. The dose was measured with a thermoluminescence dosemeter

As with therapeutic radiology, the X-ray dose should be calculated with the concepts of radiation field and volume. The skin dose is expressed in terms of absorbed dose per unit volume, which is deduced from the exposure $R$ per unit area of the skin. If the dose required for induction of skin malignancy is equal for all areas of the body, the surface-dose area integral ($R$-cm$^2$) deserves consideration. Needless to say, if the dose is high, the problem of acute dermatitis will arise; this, however, is a different problem from that of malignancy.

The skin dose varies with the magnification ratio, but the surface-dose area integral $R$ is almost similar in value to the dose from low-magnification radiography. This is also true for high-magnification radiography, provided the X-ray beam is limited so as not to project beyond the film (MAEKOSHI *et al.*, [49]) (Fig. 26).

## 3. Depth and Volume Dose

X-rays reach not only the skin but also the deep tissues in the body, so that it is necessary to account for not only the surface-dose area integral but also the volume dose, as the X-ray hazard will apply to all exposed cells.

If we compare the isodose curves of magnification radiography at fairly high and high magnification, e.g. 4 and 8 × magnification radiography, with those for roentgeno-

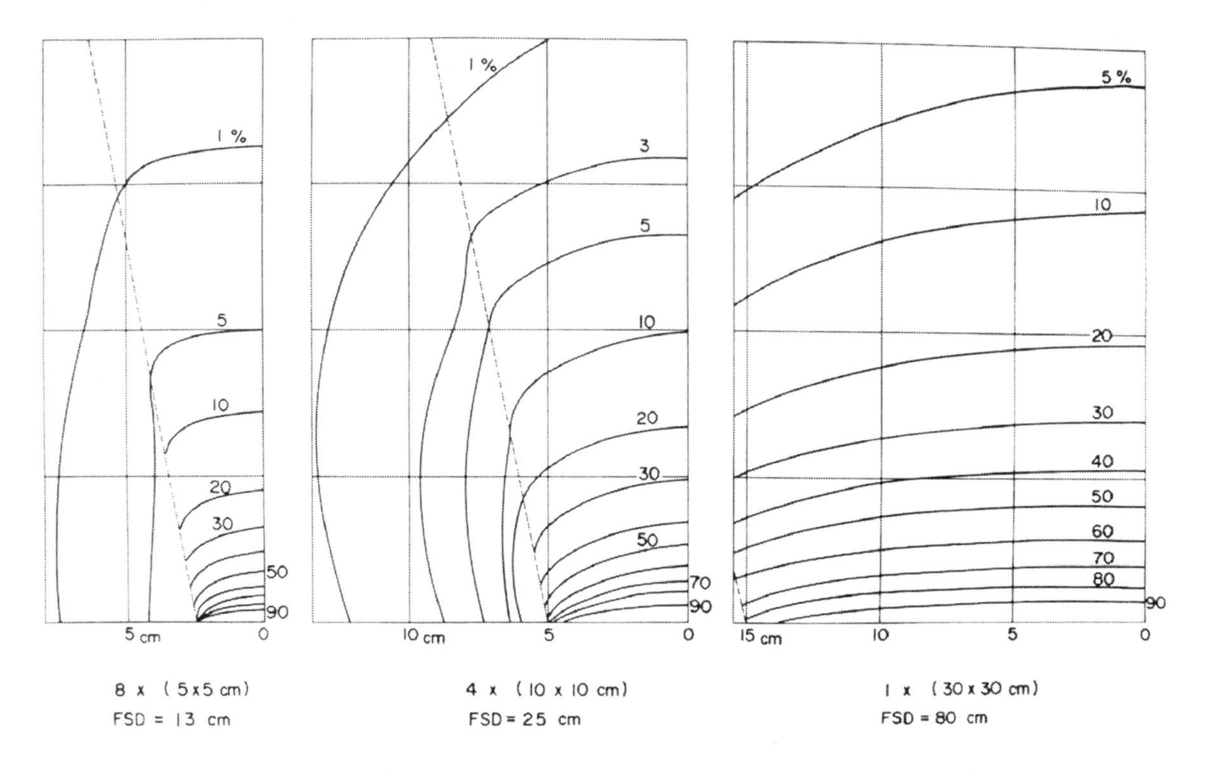


Fig. 27. Isodose curves. *Left:* 8 × magnification radiography with focal spot-skin distance 13 cm, FFD 100 cm, and radiation field 5 × 5 cm. *Center:* 4 × magnification radiography with focal spot-skin distance 25 cm, FFD 100 cm, and radiation field 10 × 10 cm. *Right:* Normal roentgenography with focal spot-skin distance 80 cm, FFD 100 cm, radiation field 30 × 30 cm, and grid of grid ratio 8. A tube voltage of 120 kV was used with a water phantom 20 cm thick

graphy with FFD 100 cm (Fig. 27), it is obvious that the higher the ratio of magnification the narrower the X-ray beam and the greater the X-ray attenuation. The fall in X-ray dose at a definite depth is in inverse proportion to the magnification ratio.

The skin dose at 8 × magnification is 4 to 5 times greater than at 4 × magnification, but the volume doses at various depths of the body are similar at 4 × and 8 × magnification within a range of difference of ±5% [120]. In normal roentgenography with grid both volume and skin doses are somewhat lower than those in magnification

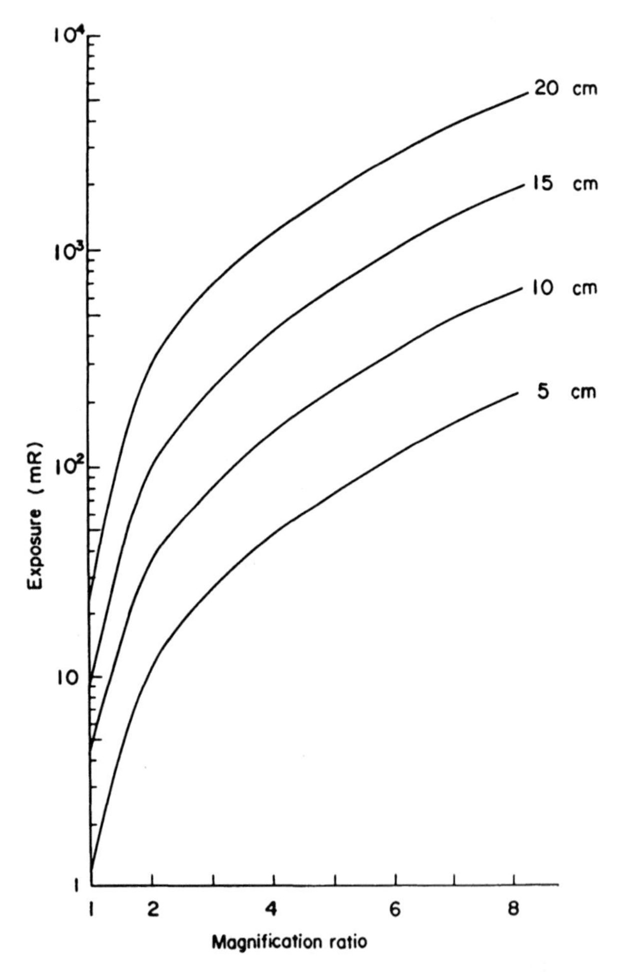


Fig. 28. Steeply increasing radiation exposure of patient in magnification radiography with increasing magnification ratio and body thickness.

radiography. If normal roentgenography is conducted with a grid of appropriate grid ratio for a body 20 cm thick, the roentgenogram will show good contrast without fog, although the exposure time will be trebled.

In sum, despite the extreme increase in skin dose in higher-magnification radiography as compared with normal roentgenography, little difference in volume dose was found between very high (8 ×) and fairly high (4 ×) magnification, provided the X-ray beam was matched to the area of the film. Volume doses were about twice those recorded in normal roentgenography. These findings, make it clear that, from the point of view of protection of the patient, when magnification radiography is conducted with an appropriately reduced radiation field, the dose computed for a given area or volume will not be large, although it may be large per unit area of the patient's skin.

## 4. Body Thickness

As in normal roentgenography, the exposure dose increases rapidly when the part to be examined is thick. Water phantoms 5 cm, 10 cm, 15 cm or 20 cm thick were exposed with a distance of 1 m between X-ray tube and film, and the dose to the radiation field of the phantom was measured with magnification ratios of 4 and 7.7 ×. With each 5-cm increased in thickness of the water phantom the radiation exposure increased approximately three to four times. It is concluded, therefore, that in magnification radiography the exposure dose increases steeply with the thickness of the part to be examined (Fig. 28).

Bodythickness also influences the magnification ratio. If a moving lesion in the thick part of the body is the object of magnification radiography, the magnification

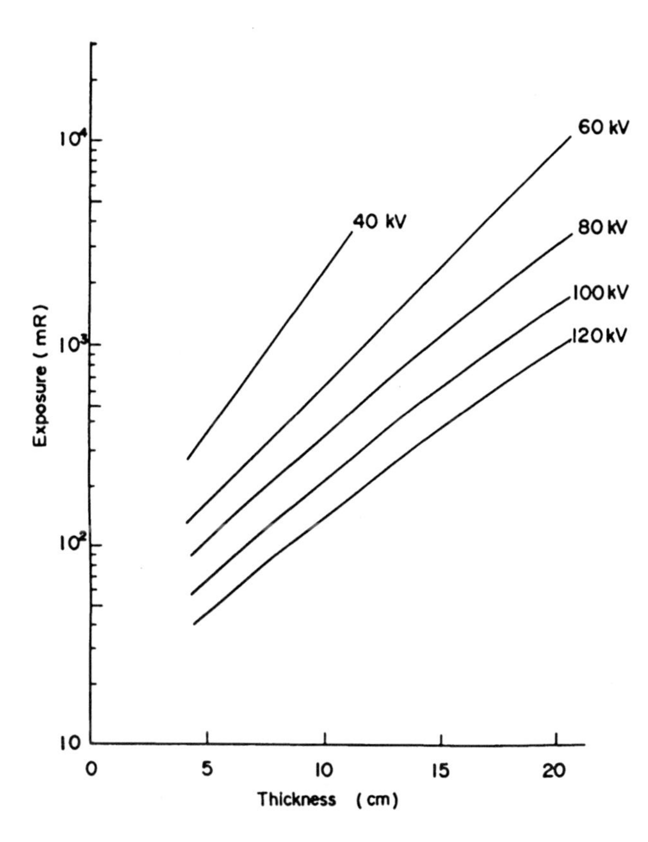


Fig. 29. Relationship between radiation dose to the skin and tube voltage in 4 × magnification radiography. Dose at various body thicknesses was measured for five different tube voltages. Focal spot-skin distance was 25 cm at fixed FFD 100 cm

ratio will be lower because a shorter exposure time is necessary. The skin dose will thus be lower than in high-magnification radiography. Nevertheless, even in relatively low-magnification radiography, the skin dose must still be considered high.

## 5. High-Voltage Technique

As in normal roentgenography, the exposure dose is lower in magnification radiography with the high-voltage technique than with normal voltage. In both cases the dose will depend on the thickness of the part to be radiographed, since the dose reaching the film rises as the third or fourth power of the tube voltage.

With a fixed FFD the dose to the skin facing the tube when the body was of a certain thickness was measured at tube voltages of 40, 60, 80, 100 and 120 kV. It was found that exposure increases rapidly as the tube voltage is lowered (Fig. 29).

Even when the high-voltage technique is used in magnification radiography, the air dose also increases rapidly with increasing magnification ratio. However, at a given magnification ratio the dose is much less than with the low-voltage technique. From the point of view of protecting the patient, the high-voltage technique thus contributes greatly to dose reduction.

However, the contrast of magnification radiographs taken by the high-voltage technique is poor. The higher the voltage, the poorer the contrast. As the tissue of interest tends to be small and thin in magnification radiography, the low-voltage technique should produce the best image. If we are ignore patient dose reduction, the low-voltage technique would be preferable. However, the high-voltage technique should be used to minimize the X-ray dose to the patient. Since scattering is negligible in magnification radiography due to the air gap between patient and film, the image quality is better than in normal roentgenography.

## 6. Patient Dose Reduction in General

As regards the genetically significant dose received, X-ray examination scores highest among man-made radiation sources. Unnecessary irradiation should therefore be

Table 2. Exposure conditions and skin dose in magnification radiography

| Part of body to be exam. | Focal spot size (mm) | Exposure factors | | | | | | | Thickness of body part (cm) | Magnification ratio relative to lesion | Maximum Magnification ratio relative to skin dose | Skin dose (rad) |
|---|---|---|---|---|---|---|---|---|---|---|---|---|
| | | view | pkV | mA | sec | (cm)[a] | (cm)[b] | (cm)[c] | | | | |
| Skull | 0.1 | lat. | 100 | 20 | 0.2 | 27 | 7 | 100 | 14 | 2.9 | 3.7 | 0.6 |
| Skull (CAG) | 0.1 | lat. | 100 | 30 | 0.15 | 27 | 8 | 100 | 15 | 2.9 | 3.7 | 0.68 |
| Sella turcica | 0.05 | a.p. | 120 | 3 | 0.43 (0.24) | 15 | 8 | 92 (69) | 20 | 4 (3) | 6.1 (5) | 1.83 (0.84) |
| Sella turcica | 0.05 | lat. | 120 | 3 | 0.21 | 15 | 8 | 92 | 16 | 4 | 6 | 0.89 |
| Neck (VAG) | 0.05 | p.a. | 120 | 3 | 0.12 | 22 | 10 | 110 | 19 | 3.4 | 5 | 0.18 |
| Lung | 0.05 | a.p. | 100 | 3 | 0.07 | 15 | 10 | 100 | 18 | 4 | 6.7 | 0.3 |
| Lung | 0.05 | p.a. | 120 | 3 | 0.04 | 25 | 9 | 100 | 17 | 3 | 4 | 0.04 |
| Lung | 0.1 | p.a. | 100 | 20 | 0.02 | 27 | 10 | 100 | 20 | 2.7 | 3.7 | 0.06 |
| Bronchial angio. or internal mammary | 0.05 | p.a. | 120 | 3 | 0.08 | 22 | 10 | 110 | 20 | 3.4 | 5 | 0.15 |
| or bronchogr. | 0.1 | p.a. | 100 | 10 | 0.05 | 27 | 10 | 100 | 20 | 2.7 | 3.7 | 0.08 |
| Bronchogr. | 0.05 | p.a. or a.p. | 120 | 3 | 0.08 | 22 | 10 | 110 | 20 | 3.4 | 5 | 0.15 |
| Dorsal spine | 0.05 | p.a. | 120 | 3 | 0.38 (0.21) | 15 | 8 | 92 | 20 | 4 (3) | 6 (4.6) | 1.55 (0.89) |
| Dorsal spine | 0.05 | lat. | 120 | 3 | 0.07 | 22 | 14 | 110 | 28 | 3 | 5 | 0.1 |
| Abdomen stomach (double contrast) | 0.05 | p.a. | 120 | 3 | 0.25 | 22 | 7 | 110 | 14 | 4 | 5 | 0.4 |
| (double contrast) | 0.05 | p.a. | 120 | 3 | 0.3 | 22 | 9 | 110 | 18 | 3.5 | 5 | 0.45 |
| (double contrast) | 0.1 | a.p. | 120 | 15 | 0.1 | 30 | 8 | 120 | 18 | 3.2 | 4 | 0.25 |
| small intestine | 0.05 | a.p. | 120 | 3 | 0.25 | 15 | 8 | 92 | 17 | 4 | 6 | 0.89 |
| Celiac angio. | 0.1 | a.p. | 120 | 15 | 0.1 | 30 | 8 | 120 | 17 | 3.2 | 4 | 0.25 |
| Sup. mesenteric angio. | 0.1 | a.p. | 120 | 15 | 0.12 | 30 | 9 | 120 | 19 | 3.1 | 4 | 0.3 |
| Inf. mesenteric angio. | 0.1 | a.p. | 120 | 15 | 0.1 | 30 | 9 | 120 | 18 | 3.1 | 4 | 0.25 |
| Renal angio. | 0.1 | a.p. | 120 | 15 | 0.1 | 30 | 7 | 120 | 19 | 3.3 | 4 | 0.25 |
| Renal angio. | 0.05 | p.a. | 120 | 3 | 0.25 | 22 | 7 | 110 | 15 (compression used) | 3.5 | 5 | 0.4 |
| Lumbar spine (II) | 0.05 | a.p. | 120 | 3 | 0.15 | 15 | 10 | 75 | 16 | 3 | 5 | 0.68 |
| Lumbar spine (II) | 0.05 | lat. | 120 | 3 | 0.8 | 15 | 13 | 84 | 25 | 3 | 6 | 3.29 |
| Lumbar spine (II) | 0.1 | lat. | 120 | 15 | 0.5 | 33 | 13 | 100 | 26 | 2.2 | 3 | 0.85 |
| Lumbar spine (II) | 0.3 | lat. | 120 | 30 | 0.5 | 40 | 13 | 114 | 26 | 2.1 | 2.8 | 0.85 |

Table 2 (continued)

| Part of body to be exam. | Focal spot size (mm) | Exposure factors | | | | | | | Thickness of body part (cm) | Magnification ratio relative to lesion | Maximum magnification ratio relative to skin dose | Skin dose (rad) |
|---|---|---|---|---|---|---|---|---|---|---|---|---|
| | | view | pkV | mA | sec | (cm)[a] | (cm)[b] | (cm)[c] | | | | |
| Head of humerus | 0.05 | p.a. | 120 | 3 | 0.07 | 15 | 6 | 126 | 10 | 6 | 8.4 | 0.36 |
| Head of radius | 0.05 | p.a. | 100 | 3 | 0.06 | 15 | 3 | 108 | 6 | 6 | 7.2 | 0.27 |
| 5th distal phalanx of the hand | 0.05 | p.a. | 80 | 3 | 0.01 | 15 | 0.5 | 93 | 1 | 6 | 6.2 | 0.04 |
| Thigh (lymph.) | 0.05 | p.a. | 120 | 3 | 0.08 | 22 | 8 | 110 | 16 | 3.7 | 5 | 0.15 |
| Shaft of fibula | 0.05 | p.a. | 100 | 3 | 0.09 | 15 | 4 | 114 | 8 | 6 | 7.6 | 0.39 |
| Medial condyle of femur | 0.05 | p.a. | 120 | 3 | 0.08 | 15 | 6 | 126 | 11 | 6 | 8.4 | 0.41 |

[a] Focal spot–table distance.
[b] Table–object distance.
[c] Focal spot–film distance.
Figures in parentheses show the reasonable skin dose obtainable by reducing the magnification ratio.

avoided. There are, however, some reasons that justify the use of magnification radiography because of the medical benefit expected from its diagnostic procedures.

The Recommendations of the International Commission on Radiological Protection, ICRP Publication 15 (1969); *Protection against Ionizing Radiation from External Sources*, state in § 156: "Provided that a certain diagnostic *procedure* as such is generally considered justified because the medical benefit to the patients is believed to outweigh the risk from the exposure, the remaining judgement in the *individual* case would relate to the choice of X-ray examination as being appropriate for that individual, the conduct of the examination, and the interpretation of the result. If any one of these three actions is impaired by ignorance, negligence or lack of resources, the radiation exposure may still be justified from the point of view of the diagnostic information but the diagnostic yield has then been obtained at an unnecessarily high cost in terms of radiation dose and corresponding risk."

§ 157: "The decision as to whether a certain radiation dose to a patient is justified is sometimes the responsibility of the referring physician, sometimes of the radiologist. In either case, however, it is imperative that the decision is based upon a correct assessment of the indications for the examination, the expected yield from the examination and the way in which the results are likely to influence the diagnosis and subsequent medical care of the patient. It is equally important that this assessment is made against a background of adequate knowledge of the physical properties and the biological effects of ionizing radiation."

§ 158: "No person shall operate radiological equipment without adequate technical competence nor apply radiological procedures without adequate knowledge of the physical properties and harmful effects of ionizing radiation."

By means of a favorable combination of high-speed intensifying screens and film, the high-voltage technique, and the appropriate magnification ratio for the thickness of the body, it is possible to keep the dose below 1 R clinically, or 2 R maximum in special cases (Table 2). The dose of 1 R more or less corresponds to the X-ray dose of a normal roentgenography of the lower abdomen. In magnification radiography it should always be kept in mind that the patient is apt to be exposed to a high skin dose.

The patient should be carefully treated and not exposed unnecessarily. For example, if 3 × magnification radiography is carelessly conducted with the combination of standard-speed intensifying screens and X-ray film at 80 kVp and a focal spot-film distance of 120 cm, the skin dose can rise to about 25 R for a serial celiac angiography involving 5 films.

It is advisable to use the high-speed screen-film combination with a voltage of 120 kV, or at least 100 kV, and to maintain the distance between focal spot and patient at not less than 15 cm, even when high-magnification radiography is carried out. A voltage of 80 kV or less is used only for very thin parts of the body. Mass surveys should not be carried out by means of magnification radiography, especially at low voltage.

When magnification radiographs are made of tissues and lesions located in the center of very thick parts of the body, the skin dose may rise to more than 1 R. In such a case the dose should decrease below 1 R, so the magnification ratio should be sacrified by shortening the distance between tube and film (Table 2).

Before magnification radiography is applied, the location of the suspected lesion will have been learned from the history of the current illness or from a normal roentgenogram taken previously. Hence, the radiation field can be kept to the minimum as the part to be examined will be small. It is advisable to keep the radiation field as small as possible. When a small field size is used for high magnification radiography, the X-ray beam should just cover the film. The direction of the X-ray beam should be chosen to avoid a radiosensitive organ or a gonad. Appropriate positioning by means of X-ray television will facilitate this procedure.

# *Chapter II.* Technique of Magnification Radiography in Clinical Practice

Magnification requires either a rather simple technique or a sophisticated one, depending on the magnification ratio. For 2× magnification radiography the X-ray unit is similar to that employed for normal roentgenography. The technique is also the same. With high-magnification radiography, however, special equipment and techniques become necessary.

## A. Equipment for High-Magnification Radiography

An X-ray tube with a small focal spot is at present available for magnification radiography. This tube can be clinically applied with an existing radiographic table. However, several types of equipment have been

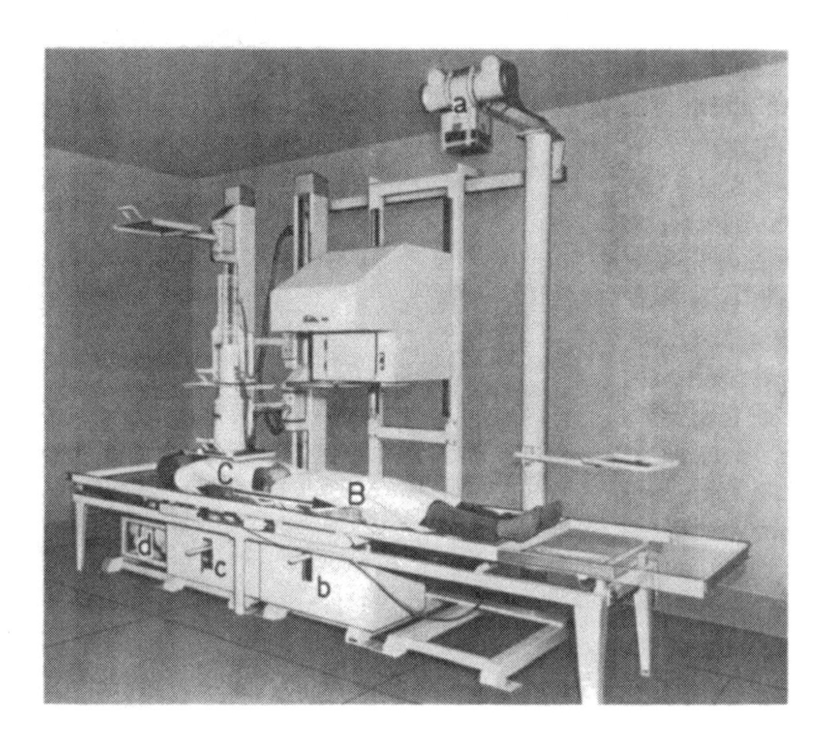


Fig. 30. Universal magnification radiography unit (Toshiba) in action. *a* X-ray tube with 2 mm focal spot for normal roentgenography. *b* X-ray tube with 0.05 mm focal spot for stationary or serial magnification radiography. *c* X-ray tube with 0.3 mm focal spot for fluoroscopy of field determination. *d* X-ray tube with 0.1 mm focal spot for magnification radiography. *C* Patient undergoing fluoroscopic examination with X-ray television. *B* Patient is moved from position C for fluoroscopy to position B for magnification radiography. This unit is not suitable for magnification angiography

specially manufactured for magnification radiography and are now available. These facilitate positioning of the patient and reduce the radiation exposure.

## 1. Magnification Radiography Unit for General Use

The equipment has a radiographic table that can be moved smoothly and accurately along the axis of the human body as well as perpendicular to it (Fig. 30). Under the table are arranged three tubes with foci of 0.1 mm, 0.3 mm and 0.05 mm respectively. Over the table is located a rotating-anode tube with a 1.5-mm focal spot for normal roentgenography. Below the table are a grid and the cassette.

For the 0.1 mm focal-spot tube, the distance between the tube and the radiographic table can be varied from 25 to 35 cm. The 0.3 mm focal spot tube is used mainly for fluoroscopy. The central X-rays are arranged to strike the center of the fluorescent screen of the image intensifier located above the table. The image intensifier is connected with an X-ray television monitor.

The 0.05 mm focal-spot tube is positioned so that the distance between its focal spot and the radiographic table is as short as 15 cm. Above the radiographic table is a film box capable of changing the cassette at least twice per second.

The central X-rays of all the X-ray tubes always hit a point on the long axis of the radiographic table.

The patient is first positioned with the aid of the X-ray television. With the 0.3 mm focal-spot tube, fluoroscopy is carried out with the television monitor and the part to be radiographed is adjusted to the center of the radiation field by shifting the table to and fro. On completion of this procedure the radiographic table is moved smoothly through the required distance, taking care that the patient does not move, so that the part to be radiographed is positioned directly above the 0.05 mm or 0.1 mm focal-spot tube.

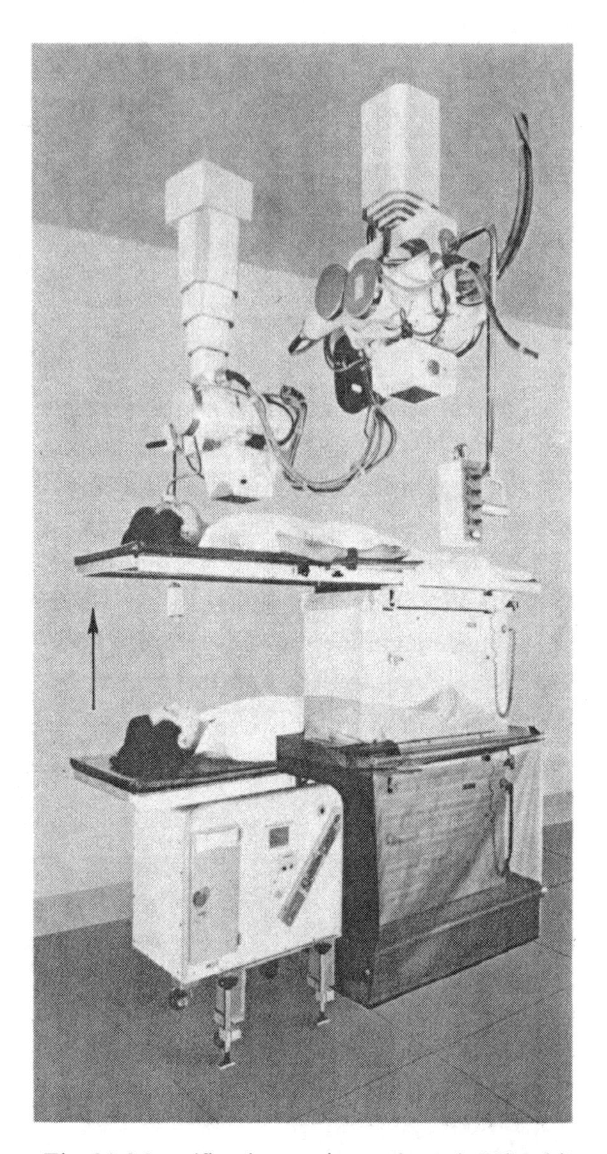

Fig. 31. Magnification angiography unit (Hitachi) in action. The area to be magnified is determined by means of X-ray television attached to the 0.6 mm focal spot tube. The table is raised close to the X-ray tube and magnification radiography is carried out with the 0.1 mm focal spot

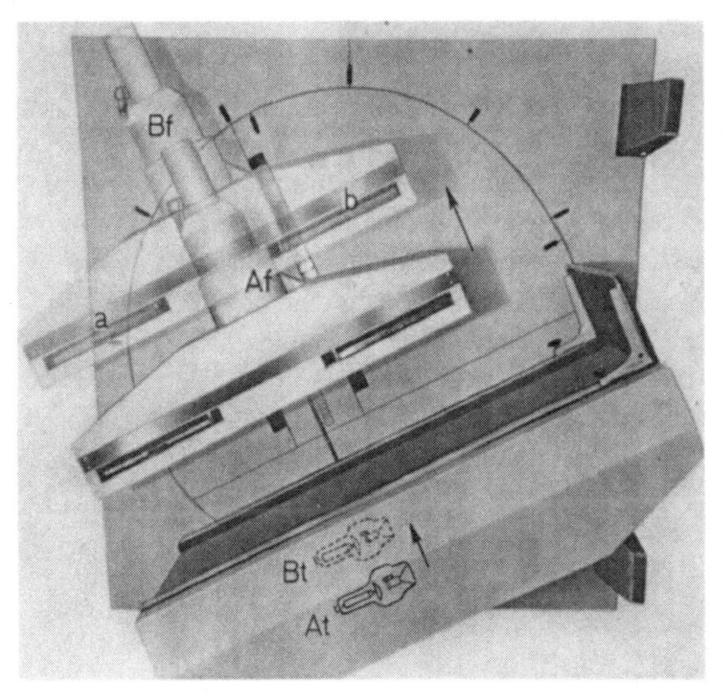


Fig. 32. X-ray television fluoroscopic unit (Toshiba) for magnification radiography of GI tract, bronchus, etc. *At* X-ray tube for fluoroscopy. *Af* Image intensifier for fluoroscopy. *Bt* Position of X-ray tube changed for magnification radiography. *Bf* Image intensifier for magnification radiography for determination of field. *a, b* Cassettes for stereoroentgenography

## 2. Magnification Angiography Unit

The equipment of an ordinary angiographic apparatus is attached to a 0.1 mm focal-spot tube.

In contrast to the magnification radiograph shown in Fig. 30, the equipment has the X-ray tube above the table (Fig. 31). Positioning of the patient is easy when the table is low. Correct fixation of the tube will minimize the blurring of the X-ray image due to vibration caused by the rotation of the anode in the tube. Good results are obtained with this unit as the 0.1 mm focal-spot tube is suspended from the ceiling with perfect fixation. The usual AOT Elema-Schönander film changer is used as it is. After normal angiography the table is moved upward to reduce the distance between the focal spot and the patient and lengthen the distance between the table and the film. A film changer such as the Siemens Puck type, if used for the ordinary angiographic table, will also be convenient for magnification angiography.

## 3. Fluoroscopy Unit Attached to Magnification Radiography Adaptor

This unit consists of a magnification radiographic tube with double focal spots attached to a fluoroscopy table. The very small focal spot is not used for fluoroscopy but solely for magnification radiography. Fluoroscopy is conducted with the large focal spot. The fluoroscopy table is of the rotary type and enables radiographs of the body to be taken at various angles, as required (Fig. 32). The apparatus is used for magnification radiography of the GI tract, bronchus, etc. The patient is examined by X-ray television with the X-ray tube 45 cm from the table.

Once the part of the body to be radiographed has been determined, the table is raised to give an X-ray tube-table distance of 15 cm. The image intensifier is located away from the central X-ray, the top of the cassette holder is adjusted to the central X-ray and radiography is conducted.

However, magnification radiography units without an X-ray television set attached are more widely used at present. With this type of apparatus the patient is positioned by the techniques described in Chap. II.C.2.

## B. Accessories for Magnification Radiography Unit

### 1. Filter and Cone

A filter is placed at the radiation mouth to provide the total filtration equivalent to 2.5 mm Al, which is necessary to protect the patient.

In order to reduce the size of the X-ray beam as required, it is useful to have a cone or collimator, especially a multileaf collimator with light projector, to indicate the radiation field.

If we wish to obtain a 2× magnification radiograph, the radiation field on the skin of the patient needs to be reduced to approximately one half of the field used for normal roentgenography, where the X-ray beam is adjusted to the size of the film. The reduction procedure ensures that the X-ray beam does not project beyond the X-ray film. As the radiation field and the light field sometimes do not coincide, regular checking is needed. For 6 or 8× magnification radiography the field will be extremely small. The size of this field for 6× magnification is 5×5 cm; this will completely cover an X-ray film 35×35 cm. On occasion, the field is reduced to 2×2 cm; the field size will then be enlarged to 12×12 cm in case of 6× magnification. Especially in such a case, television monitoring is needed to ensure correct positioning of the X-ray beam. In magnification radiography the X-ray tube is placed under the table, therefore the light projector is useless. It is then necessary to use the X-ray television for correct positioning as well.

### 2. Grid

A grid is usually not used in magnification radiography, as the air gap between the patient and film eliminates the need for the grid (MOORE [52]). According to HALE and MISHIKIN [34] the air gap in 2× magnification radiography corresponds to a grid with grid ratio of 15. Despite this, however, it has been found experimentally that the use of the grid can improve the quality of the magnification radiogram (SANDOR *et al.* [84]). If a grid is used, however, two or three times longer exposure time is needed (ZIMMER [140]; VAN DER PLAATES [61]; TAKAHASHI and KOMIYAMA [102] and others); for this reason the grid is not normally used.

### 3. Intensifying Screens and Film

In normal roentgenography the extent of blurring of an object is determined by the nature of the intensifying screens. At present, intensifying ratios of 40, 30 or 20 represent high-speed (HS), medium-speed (MS) and fine-grain (FS) screens, respectively; each is the trade mark of Kyokko (Dai Nippon

Toryo Co, Japan) and will resolve 0.16, 0.2 mm, 0.09, 0.1 mm or 0.06 mm of a test object placed directly on the surface of the cassette. X-ray images smaller than the above will thus be blurred out. Exposure would be greatly reduced if the newly developed rare earth-oxysulphride screens or -phosphosulphide screens are combined with suitable X-ray film and used instead of existing calcium tungustate screens. Usually, however, the object is placed distant from the cassette. The size of the focal spot may also produce additional blur of penumbra besides the blurring due to screens.

In normal roentgenography X-ray images of small lesions cannot be obtained due to lack of sharpness caused by the intensifying screens. A sharp X-ray image is obtainable in magnification radiography, even in such cases. This is due to the resolution effect, whereby the blur produced by the intensifying screen is overcome by the magnified image. This is a great merit of magnification radiography.

When magnification radiography is conducted with a 0.05 mm focal-spot tube and MS screens, the maximum size of the image blurred out will be 0.13 mm in 2 × magnification and 0.04 mm in 4 × magnification.

In order to reduce blurring to a minimum it may be advisable to use a fine-grain screen. In the early days of research on magnification radiography it used to be said that a fine-grain intensifying screen should be used, but today it is generally accepted that there is no need for a specially fine-grain screen.

The use of a fine-grain film, without an intensifying screen, allows extremely small objects to be imaged by optical magnification, but in clinical practice intensifying screens are indispensable to shorten the exposure time.

Intensifying screens of too high speed are not appropriate either. In such high-speed screens the grains are rough, and this affects the quality of the roentgenogram, even in magnification radiography. Hence, in practice medium-speed (MS) intensifying screens are used. Recently high-speed screens with good sharpness have become available. The use of intensifying screens also improves contrast and this assists identification of the image.

The film should be of as high a speed as possible; the use of films rich in grain or mottling should be avoided. Any high-speed X-ray film is suitable, and there is at present no X-ray film that is specific for magnification radiography. We use Kodak RPR, Fuji KX or Sakura QX.

## C. Positioning

If the lesion to be radiographed is situated deep within the body, and the lesion is not equidistant from the skin of the front and back of the body, the patient is positioned with the lesion as near as possible to the X-ray tube, thus making the magnification ratio as high as possible. The prone or supine position on the roentgenographic table is chosen on this principle.

In high-magnification radiography the central X-rays should precisely strike the tissue to be examined when the patient is lying on the table. In normal roentgenography or low-magnification radiography, the part to be examined usually does not extend beyond the film even if the central X-rays are slightly off target. In high magnification radiography, however, such extension or partial coverage of the film by the X-ray beam can easily take place. Careful positioning of the patient before magnification radiography is therefore essential.

## 1. Positioning Based on Anatomical Knowledge

Positioning requires a sound knowledge of roentgen anatomy. Some identifying anatomical marks that are palpable or recognizable externally, such as projection of a bone, fissure, meatus or pit on the face or trunk, can help in the location of a small tissue or organ in the body (Ono *et al.* [59]).

In normal roentgenography, a general roentgenologic view of the body is obtained without regard to the position of the image of the lesion on the roentgenogram. In magnification radiography, the central X-rays must hit a small tissue or lesion. Such precision is difficult to achieve unless strict attention is paid to prearranged procedures because in magnification radiography the distance between focal spot and object is shorter and that between body and film longer than in normal roentgenography.

As film of $30 \times 25$ cm is generally used, the radiation field must be made smaller to match the magnification ratio.

Anatomical knowledge is applied to make the central X-ray strike the tissue of interest. Even so, a slight misdirection of the central X-rays will produce large errors on the film and the image of the lesion may not be recorded on the film. The technique of hitting the tissue of interest is not easy.

## 2. Positioning Based on Interpretation of Previously Taken Normal Roentgenogram

The patient may be positioned on the table after study of a normal roentgenogram in which the central X-rays do not hit the lesion, although they hit the center of the film. First, the distance between the center of the roentgenogram and the image of the lesion where the central X-ray must fall is measured on the patient's skin. Instead of a routine roentgenogram taken preliminarily for the purpose of obtaining a general view, normal roentgenography with a small-sized film can be helpful.

The patient is fixed on the roentgenographic table and a normal roentgenogram is taken and developed. The part to be magnified is determined by studying this roentgenogram and is marked on the skin. The central rays of a very small focal-spot tube are adjusted to this mark and magnification radiography is then performed.

If the site of the lesion lies away from the center of the magnification radiogram, the table is moved through the necessary distance to correct the position and magnification radiography is repeated. This method is time-consuming.

## 3. Positioning by Means of X-Ray Television

The part to be radiographed is first examined by X-ray TV fluoroscopy and the tissue of interest is adjusted to the central X-rays. When the X-ray beam exactly conforms to the lesion, the radiographic table is moved as required and the central X-rays of the magnification radiography tube are focused on the part to be examined. When magnification radiography of a lesion located in a deep part of the body is to be performed, it is essential to make a preliminary exploration by X-ray TV fluoroscopy.

Positioning by X-ray TV is usually simple, but it can be difficult to use on some parts of the body because of its complicated structure, or its use may not be necessary depending on the site to be examined. Even when using X-ray TV, repetition of unnecessary fluoroscopy should be avoided once the part to be radiographed is established.

## D. Magnification Ratio

The magnification ratio to be applied varies with the part of the body to be magnified: the appropriate magnification ratio is determined in accordance with thickness of the body or the movement of the organ.

### 1. Definition of Magnification Ratio

Magnification ratio has to be considered from two points of view: clinical significance, and radiation protection.

#### a) Magnification Ratio Relative to Lesion

If the distance between the lesion to be examined and the patient's skin facing the X-ray tube is $p$, the magnification ratio will be (cf. Fig. 24):

$$\alpha_1 = \frac{a+b}{a+p}$$

where $a$ is focal spot-skin distance and $b$ is skin-film distance. The term $\alpha_1$ indicates magnification ratio relative to lesion and represents the actual magnification ratio of the lesion or tissue (Fig. 33). Tissues that are closer to the tube than the lesion will be magnified more and those further away from the tube magnified less. For calculation of

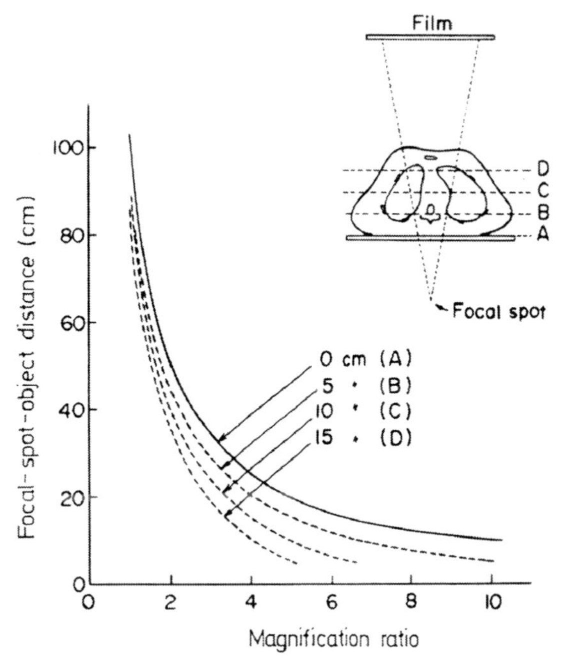


Fig. 33. Relationship between focal spot-object distance and magnification ratio. *Upper right:* Lesion is located 0 cm, 5 cm, 10 cm or 15 cm away from the skin. Magnification radiography FFD 100 cm

the magnification ratio, the value of $p$ should be accurately known. This is obtained by taking an axial transverse tomogram in the horizontal position (TAKAHASHI [117]) and measuring the distance of the lesion from the skin.

If no horizontal type of axial transverse tomographic apparatus is available, a lateral-view roentgenogram can be used for an approximation. This means that the magnification ratio for the lesion $\alpha_1$ is not usually accurate.

#### b) Magnification Ratio Relative to Skin Dose

In TAKAHASHI's earlier method, the magnification ratio was computed taking the skin facing the focal spot as the standard. Thus, when the focal spot-skin distance is $a$, and the table-film distance $b$, the magnification ratio $\alpha_p$ will be $\alpha_p = \frac{a+b}{a}$. This procedure allows the magnification ratio to be determined precisely, as $a$ and $b$ can be accurately measured (cf. Fig. 24).

However, when 4× magnification radiography of a thick part of the body such as the chest is to be done, the magnification ratio $\alpha_p$ fails to coincide with $\alpha_l$. The part of the lung near the skin will be magnified about 3.5×· while what far from the skin will be about 2.5× and hence much smaller than the magnification ratio at the skin (cf. Chap. I.D.5: Variability of magnification ratio).

From the point of view of the exposure however, it is better to use the magnification ratio $\alpha_p$. $\alpha_p$ is therefore called the magnification ratio relative to skin dose.

As stated elsewhere, when the magnification ratio is $\alpha_p$, the exposure dose at the patient's skin will be approximately

$$D \cdot \alpha_p^2.$$

When the FFD is fixed, $D$ is the approximate skin dose in normal roentgenography conducted with a grid of appropriate grid ratio, irrespective of the focal-spot-patient distance and of the thickness of the part of the body to be examined.

The protection problem should never be neglected in magnification radiography in favor of a high magnification ratio. If there is any risk of undue radiation exposure, the magnification ratio should be reduced.

The following relationship exists between $\alpha_l$ and $\alpha_p$:

$$\frac{\alpha_p}{\alpha_l} = \frac{\frac{a+b}{a}}{\frac{a+b}{a+p}} = 1 + \frac{p}{a}.$$

As the value of $\alpha_p$ can be obtained accurately, $\alpha_l$ can be calculated by the formula $\alpha_l = \frac{a}{a+p}\alpha_p$. While $\alpha_p$ is usually an integer, $\alpha_l$ is usually a fraction. The value of $\alpha_p$ is always a higher than that of $\alpha_l$. The larger the value of $p$ and the smaller that of $a$, the more $\alpha_p$ will exceed $\alpha_l$.

## 2. Determination of Magnification Ratio for Non-Moving Organ

The size of the penumbra for a given focal spot of the X-ray tube can be determined once the magnification ratio is known for a fixed distance between the focal spot and the film.

The size of the focal spot must be determined by the measurement described (cf. Chap. I.C.2) as it differs for individual tubes. One of the decisive factors in preferring the magnification ratio based on lesion, $\alpha_l$, is the size of the focal spot.

Fig. 34 shows the relationship between focal spot size, magnification ratio and resulting size of penumbra. Thus, it is possible to determine the magnification ratio from the appropriate size of penumbra. The solid curve (FFD 100 cm) and the dashed curve (FFD 150 cm) have been compiled from experiments and clinical experience.

For magnification radiography with FFD 100 cm and focal spot-lesion distance 25 cm (abscissa), the magnification ratio $\alpha_l$ can be read off from the left ordinate and the 100-cm curve (solid line). The magnification ratio is four. Next, the size of the penumbra is obtained from the curves at the top of the diagram. When the distance from the focal spot to the lesion is 25 cm, it is seen from the right ordinate that the penumbra is

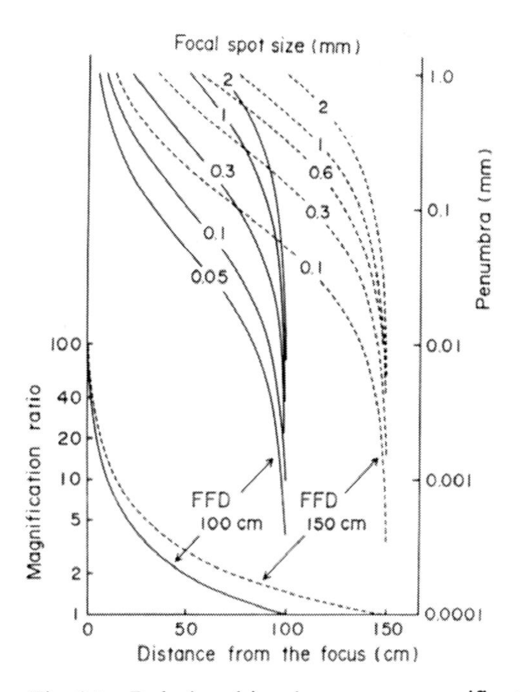


Fig. 34. Relationship between magnification ratio, size of focal spot of X-ray tube, and induced penumbra. *Solid line.* FFD 100 cm, *Dashed line.* FFD 150 cm. *Example:* 6.5 × magnification will be possible with a penumbra of less than 0.2 mm when a 0.05 mm focal-spot tube is used at focal spot-skin distance 15 cm and FFD 100 cm (solid line)

0.12 mm for the 0.05 mm focal spot size. Assuming that a penumbra up to 0.2 mm in size is not recognized as blurring, 4 × magnification radiography would provide a sharp image with the given exposure conditions.

At 6 × magnification, the solid curve for the 0.05 mm focal spot indicates a 0.2 mm penumbra, while the solid curve for 0.1 mm focal spot shows that a 0.2 mm penumbra occurs at 3.5 × magnification.

The dashed curves are similarly used for, when a magnification radiogram is to be taken at FFD 150 cm.

When the proper magnification ratio has been determined, it is next necessary to select the tube current (mAs) for magnification radiography. We reviewed good magnification radiographs taken in our Departments and determined empirically the optimal exposure conditions for magnification radiograms of soft tissue, bone, lung or bronchus, GI tract, and vessels filled with contrast medium. Magnification radiography is usually performed with a fixed FFD of 100 cm, MS intensifying screen, and a tube voltage of 120 or 100 kV. The resulting range of radiographic relationships between mAs and body thickness is shown for 120 kV in Fig. 35 B, and for 100 kV in Fig. 36 B. These figures include a zone in which conditions differ according to the physique of the patient.

For radiography of the bones and stomach mAs is greater than for soft tissues of the same thickness, because a film density of more than 1 is needed to give a good image of the bones and stomach.

For the lung, the tube current should follow the upper line for persons of muscular physique and the lower line for fatty persons. When the focal spot-skin distance is 15 cm and the focal spot-lesion distance 25 cm, 4 × magnification radiography of the lung is performed with less than 0.4 mAs at 120 kV.

With the radiation risk to the patient in mind, supplementary diagrams (Figs. 35 A and 36 A) are provided for reading off the radiation exposure of the skin. These diagrams were obtained by placing the test object at various depths of the water phantom and carrying out magnification radiography with 120 kV and 100 kV and the necessary radiation dose to give the density of 1.0 on the film.

From these figures we obtained the relationship between the thickness of the water phantom and tube current. The value of mAs for magnification radiography of lung, bone, and contrast media in the body can be derived from the value of mAs for

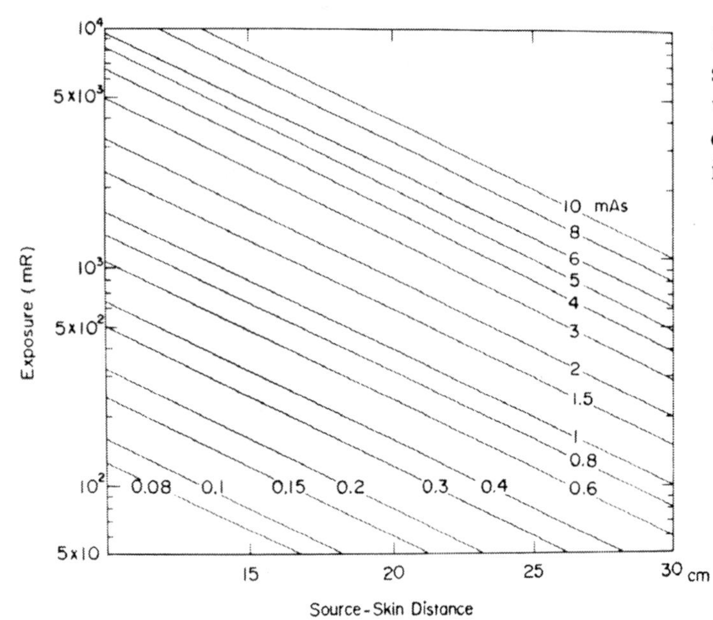


Fig. 35 A. Relationship between radiation exposure to the skin and current of tube circuit (mAs) with respect to variation of the focal spot-skin distance at fixed FFD 100 cm, tube 120 kV and image density 1.0

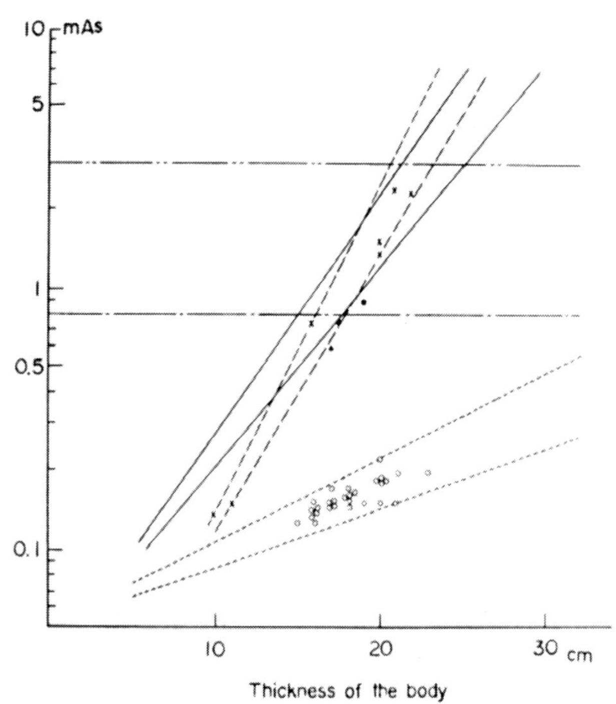


Fig. 35 B. Relationship between body thickness and mAs with tube voltage 120 kV in magnification radiography. Area between dotted lines. Exposure condition for chest ○. Area between dashed lines. Exposure condition for bone or contrast medium in the soft tissue ×. Area between solid lines. Exposure condition for water to give a density of 1.0 on the film. Horizontal line ——·——· shows exposure of 1 R at the skin of the patient when 0.05 mm focal-spot tube is used at 6.5 × magnification radiography with focal spot-skin distance 15 cm and FFD 100 cm. Horizontal line ——··——·· shows exposure of 1 R at the skin of the patient when 0.05 mm or 0.1 mm focal-spot tube is used at 4 × magnification radiography with focal spot-skin distance 25 cm and FFD 100 cm

magnification radiography with different thicknesses of water. Magnification radiography of the water phantom was carried out with FFD 100 cm at a focal spot-skin distance of 15 cm, 20 cm and 25 cm. The relationship between mAs, magnification ratio and radiation exposure at FFD 100 cm was then plotted (Figs. 35 A and 36 A).

The appropriate mAs for the magnification radiography of lung, bone or contrast media at a safe level of radiation exposure of the skin of the patient can be obtained from Figs. 35 A, B and 36 A, B. We use these figures in actual magnification radiography to determine the appropriate mAs for lesions in the parts of the body of known thickness.

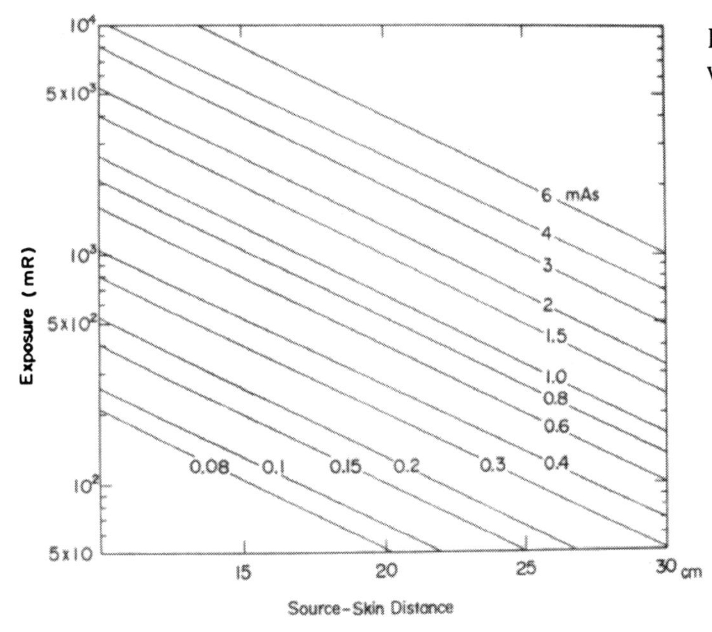


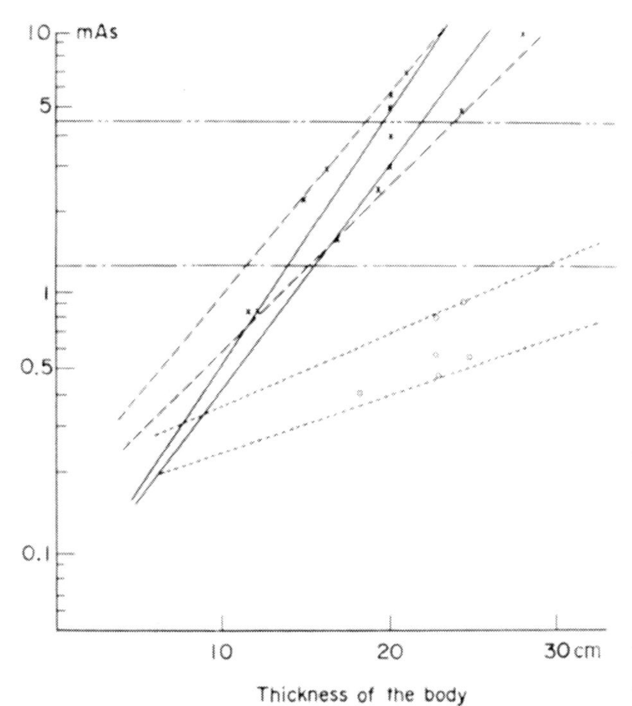


Fig. 36A and B. As Fig. 35A, B but with tube voltage 100 kV

The exposure time should be chosen so as to give a magnification radiograph with a sharp X-ray image. For a stationary organ the exposure time is fairly long when fixation of patient is perfect during magnification radiography. With FFD 100 cm and fixed mAs, the skin dose will be proportional in accordance with the inverse-square law to the magnification ratio, as represented by the focal spot-skin distance (Figs. 35A and 36A).

The tube current (mAs) needed for magnification radiography of the lung, bone, or contrast media is read off from Figs. 35B and 36B, and converted to skin dose by means of Figs. 35A and 36A. If the skin dose is to be kept below 1 R, 4 or 6× magnification will sometimes be difficult to achieve. In such cases, the magnification ratio should be reduced.

## 3. Determination of Magnification Ratio for Moving Organ

When a tissue or lesion in the body is more or less in motion, the blurring due to movement will be substantial with high-magnification radiography.

In order to obtain a sharp X-ray image on the magnification radiograph, it is necessary to consider the relationship between the grade of movement of the organ and the thickness of the body part where it is located. If the lesion is in a thick body part, the skin dose will be high. In addition, a longer exposure time is apt to be needed and lack of sharpness is induced.

The capacity of an X-ray tube with a very small focal spot is poor and this limits exposure time. This type of tube is therefore not suited for magnification radiography of a fast-moving tissue or organ.

Parts of the body differ in composition, due to the presence of air, soft tissue, or calcium. To account for these factors, the several parts of the body are represented by different thicknesses of water, as in Figs. 35 and 36.

The test object is moved in the water phantom at a known, uniform speed to represent the movement of an organ, and the magnification radiograph is taken. With the aim of obtaining a sharp image of test object, the applicability of 0.05 mm, 0.1 mm, and 0.3 mm focal-spot tubes to magnification radiography conducted at 120 kV was investigated. Fig. 37 shows the relationship between the speed of movement of an organ and body thickness where the lesion is located. The correctness of this diagram was confirmed by the series of experiments shown in Fig. 23.

The areas A, B, C or D of Fig. 37 indicate the optimal resolving image in density and sharpness for the 0.05 mm, 0.1 mm, 0.3 mm and 1.5 mm focal spot tubes, respectively, for a tissue moving at a given speed. For example, a lesion at the center of a chest 20 cm thick (which corresponds to 5 cm of water) moving at a speed of 2 mm/sec is within the capacity of a 0.05 mm focal-spot tube. A sharp magnification radiograph is obtained at

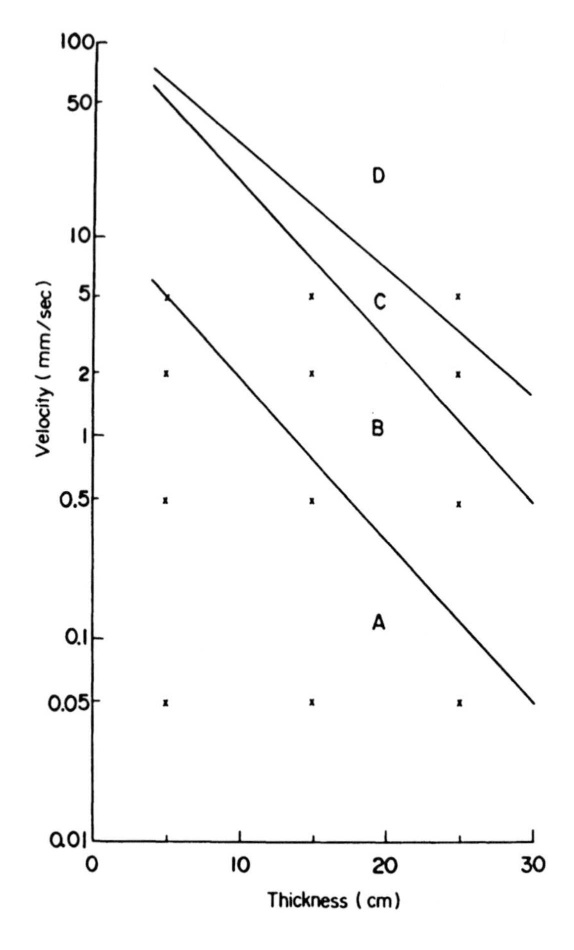


Fig. 37. Diagram showing exposure conditions for magnification radiography at 120 kV to obtain a sharp image of a moving organ in the body at FFD 100 cm. X-ray tubes with focal spots of different sizes were used. *Area A* 6× magnification radiography: sharp image with 0.05 mm focal spot tube of maximum output 3 mA at 120 kV. *Area B* 4× magnification radiography: sharp image with 0.1 mm focal spot tube of maximum output 20 mA at 120 kV. *Area C* 2× magnification radiography: sharp image with 0.3 mm focal spot tube of maximum output 30 mA at 120 kV. *Area D* normal roentgenography: sharp image with 2.0 mm focal-spot tube. × Values on the line from bottom to top at left obtained from Figs. 26, 27, 28 and 29. × Values on the line from bottom to top at center obtained from Figs. 30, 31, 32 and 33. × Values on the line from bottom to top at right obtained from Figs. 34, 35, 36, 37 or 38. The diagram shows that a magnification radiograph of the thin part of the body, even with moderate movement, can be taken with an X-ray tube with 0.05 mm focal spot

120 kV. Although a magnification radiograph of the lung taken with 0.05 mm focal-spot tube provides better image quality than one taken with 0.1 mm focal-spot tube, it is better than the normal roentgenogram and is therefore, applied also to magnification radiography of the lung (Table 2). For an object lying in an abdomen 15 cm thick and moving at a speed of 1 mm/sec, it will be necessary to use an X-ray tube of 0.1 mm focal spot to get a sharp image.

When a tube voltage of 100 kV is applied to magnification radiography of a moving organ, the area A would greatly reduce as compared with that of Fig. 37, because of the increased exposure time (cf. I.F.5. High-voltage technique). Development of rare-earth screens or X-ray tubes of a higher capacity would enable application of lower voltage to magnification radiography of moving organs. With the grid-biased focal-spot tube, reducing the size of the focal spot results in a fall of capacity due to longer exposure time. In magnification radiography of a moving organ or tissue the exposure time should be reduced, if necessary, by using a higher-capacity tube with a large focal spot.

To keep the X-ray dose reaching the film high, there is a temptation to reduce the distance between focal spot and film. However, it is not advisable to make the focal spot-skin distance less than 15 cm because the skin then comes too close to the focal spot and receives much radiation of scattered rays from the tube. Moreover, it is very difficult to shorten the distance between the focal spot and the patient's skin, because the variability of magnification ratio for tissues in the body is much increased. Thus, when the lesion is situated deep in the thick body, higher values of $\alpha_1$, the magnification ratio relative to the lesion, are limited.

## E. Field Size

The choice of radiation field depends on the condition of the patient and/or the lesion. If the lesion is believed to be located uniformly in the body or organ—for instance, silicosis or fibrosis of the lung—the field can be at any part of the suspect tissue. If the lesion is detected on interpretation of the normal roentgenogram, that lesion alone should be covered with the X-ray beam of magnification radiography.

It is a basic principle that the X-ray beams should not project beyond the area of the film. When using an 0.05 mm focal-spot X-ray tube located under the roentgenographic table and keeping the distance between focal spot and radiographic table constant, say 15 cm, with FFD 90 cm and film size 35 x 35 cm, the radiation field size is determined for the skin facing the X-ray tube. If the distance between focal spot and table is to be variable, the radiation field should be changed in accordance with the change of magnification ratio. The relationship can be studied on a diagram (see Fig. 25). At radiography the aperture at the radiation mouth is opened appropriately in accordance with the diagram.

In cases where the X-ray tube is located above the table, it is quite simple to use the appropriate multileaf cone with a light projector. Irrespective of the magnification ratio, the X-ray tube and the cassette are set so that the light beam from the tube just covers the cassette, and then the patient is positioned.

When the lesion is comparatively small with no pathologic changes in the surrounding tissue, the radiation field should be made as small as possible, especially in high magnification radiography. For example, when using a 0.05 mm focal spot tube and an FFD of 15 cm, the film can be 25 × 30 cm or less. For 8 × magnification radiography a radiation field size of 2 × 2 cm is sometimes selected. Even in such a case, the findings would be sufficient for examination. However, when the radiation field is very small, it is difficult to make the X-rays to hit the lesion unless X-ray TV is used. Even with X-ray TV, a lesion in the lung may move out of focus due to respiration. Also, if stereoscopic magnification radiography is required, there will be some difficulty in obtaining a stereoscopic image [37]. These must be considered when a very small field size is adopted.

## F. Test Marker

In magnification radiography it is useful to have a method by which the resolving power of the focal spot is indicated directly on the roentgenogram. A small test marker is placed on the radiographic table for this purpose.

The grid-biased focal spot is usually sufficiently small. Some times, there may be a risk of performing magnification radiography with a focal-spot unsuitable for high magnification; this may cause a filter effect that will prevent correct diagnosis. Simultaneous magnification radiography of a test marker is helpful in such a case.

The test marker should be small so as not to interfere with interpretation of the magnification radiograph. For example, test marker A was prepared by inserting into an acrylic resin square 10 × 10 mm 3 lengths of lead wire of diameter 0.1 mm, 0.08 mm, 0.06 mm and 0.04 mm, arranged parallel at intervals equal to their diameter (see Figs. 47 B, 53 C, and 54 B). Even with such small objects this marker is imaged for thin parts of the body, but its image does not interfere with interpretation. Test marker B is made of gold foil 40 mm thick, with 3 lengths of wire of diameters 0.03 mm, 0.043 mm, 0.062 mm and 0.09 mm arranged parallel at intervals equal to their diameters (Fig. 38, bottom left). This was also set in acrylic resin 20 × 20 × 1 mm.

The test marker is placed away from the center of the radiation field. With marker A it is better to use two at right angles to each other (see Fig. 53 C).

A 0.05 mm focal-spot tube intrinsically has the ability to resolve objects of 0.025 mm, but in practice it can barely resolve a wire of 0.04 mm. This is because scattering occurs in the human body and lowers the image quality. If the 0.04 mm wire is resolved, the focal spot can be considered of appropriate size for 4 or 6 × magnification radiography (Fig. 38, left and right).

However, the magnification ratio obtained by this method refers to the skin dose, $\alpha_p$, and not to the magnification ratio relative to the lesion, $\alpha_l$. Hence, it is not possible to equate the magnification ratio based on the image of the test marker with its actual size. Furthermore, the resolution of the test marker does not take account of the blur due to motion. Even when a sharp image of the test markers is obtained, the lesion, if in motion, will not be sharply imaged.

## G. Exposure Factors

High-magnification radiography is carried out under the radiographic conditions set out in Table 2. Note that a 0.05 mm, 0.1 mm, or 0.3 mm focal spot tube can be safely used with tube currents of 3 mA, 20 mA and 40 mA per 0.1 sec at 100 or 120 kV,

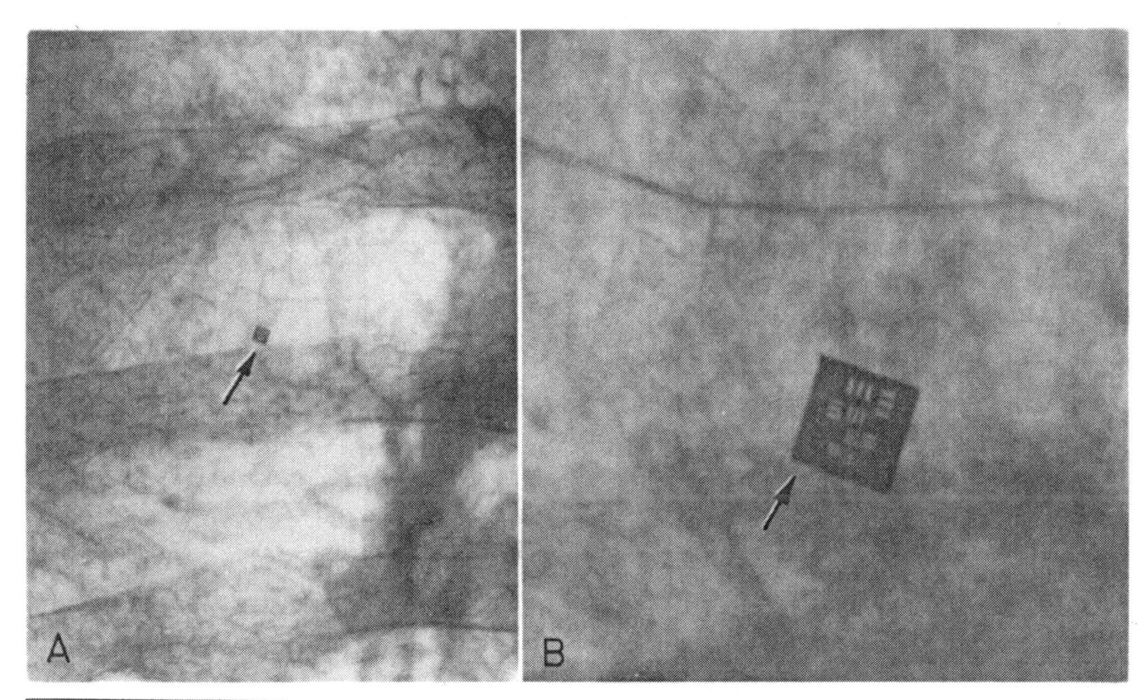


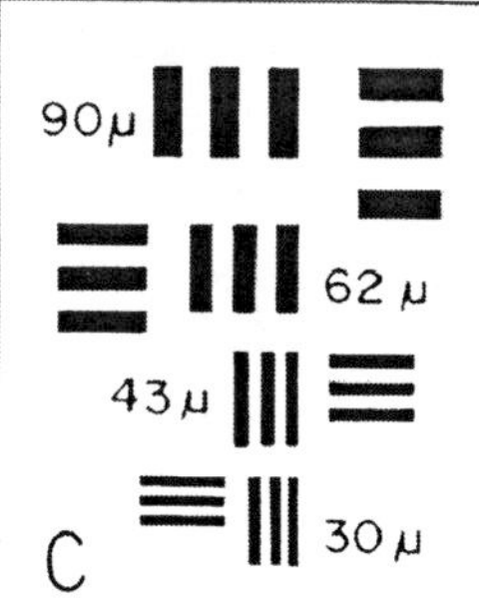


Fig. 38. Image of test marker. Test marker used with magnification radiography of chest. *Bottom left:* Diagram of test marker. *Top left:* Normal roentgenogram of subject lying on the roentgenographic table; test marker (↗) placed on table top. FFD is 200 cm. Test marker is not resolved by naked eye; 0.9 mm wires are resolved when magnified optically. *Right:* 6 × magnification radiograph taken with FFD 100 cm. 0.043 wires of test marker are resolved clearly along the body axis and, though somewhat indistinctly, also perpendicular to the body axis

respectively. Low-voltage technique should not be used for radiography of bones or contrast media, even though contrast is improved at low kVp. When body thickness differs from the norms in Table 2 (cf. p. 44), adjust by 2 kV for every 1 cm difference in thickness. The X-ray output will vary in proportion to the mAs: for a body thicker than the standard man in Table 2 (cf. p. 44), the mAs is increased by 25% for each additional thickness of 1 cm.

Another way to control exposure is by changing the tube current. With increasing focal-spot size from 0.05 mm, to 0.1 mm, to 0.3 mm, the X-ray dose rate decreases in this order even when the same mA is used. The maximum tube current increases considerably when the focal spot becomes large. It is therefore advisable to use a 0.1 mm focal-spot tube rather 0.05 mm when the exposure time has to be shortened.

For 6 × magnification radiography of the bones, a long exposure time may be used when the patient is firmly fixed.

For angiography or radiography of the digestive tract the exposure time has to be short, and for 3 to 4 × magnification radiography of the trunk in particular, a 0.1 mm focal-spot tube is appropriate. In high-magnification radiography due attention should always be paid to any increase in exposure time and skin dose.

## H. X-Ray Television Magnification Fluoroscopy

Normal fluoroscopy employing a grid-biased focal spot tube is considered here. However, with the combined use of X-ray TV and magnification fluoroscopy with the tube employed for magnification radiography, the patient's skin dose can be kept relatively small, and dynamic observation of minute changes of the moving lesion becomes possible.

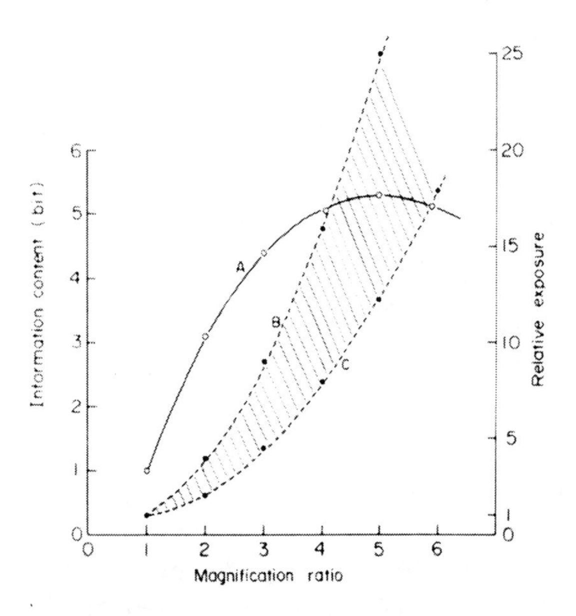


Fig. 39. Information content and radiation exposure in magnification fluoroscopy. 2× magnification is made either by the electron-beam method or the geometric one, 3× magnification is effected by 2× electron magnification and 1.5× geometric magnification, and 4× magnification by 2× electron and 2× geometric magnification. The information content increases with magnification ratio but becomes saturated beyond 4× magnification fluoroscopy. Radiation exposure, represented the hatched zone, increases with magnification ratio. The upper limit of the zone represents a thick object and the lower limit a very thin object

There are two ways of performing such fluoroscopy: (1) the patient is positioned between the tube and the image intensifier, and TV fluoroscopy is conducted in geometrical magnification; (2) during the procedure of regular TV fluoroscopy, the electron beam is biased within the image intensifier and a magnified image is obtained on the TV monitor screen.

The two methods showed no difference in the quality of the magnified image when a focal spot tube of 0.3 mm was used (FUJITA [30]; STARGARDT [95]). In X-ray TV magnification fluoroscopy, image quality improves as the geometrical magnification increases to 2, 3 and 4×, but not beyond. In electron-beam magnification, image quality improves up to 2× magnification and deteriorates thereafter. The reason why image quality is improved by magnification to 2× by electron-beam magnification and an additional 2× by the geometrical magnification method is that the blurring due to the fluorescent screen of the image intensifier and that of the TV monitor becomes negligible when the X-ray image is magnified.

The skin dose of X-rays due to 2× magnification by the electron beam method is increased approximately two- or threefold. The 4× magnification achieved by combining 2× geometrical amounts to about a fourfold magnification is thus necessary to conduct magnification fluoroscopy under conditions where the image quality is good and the skin dosage low. If 2× magnification is the object of fluoroscopy, the electron-beam magnification method meets the need because of its simplicity. For 4× magnification, electron-beam fluoroscopy in combination with 2× geometrical magnification is recommended, as by this method the skin dosage can be limited to about 6 R per minute.

X-ray TV magnification fluoroscopy should in principle be conducted with due care in the clinic, and is not generally recommended. According to ICRP, fluoroscopy should be used only for dynamic studies. At the present time the need for a dynamic study by magnification fluoroscopy is far less than that by conventional fluoroscopy. We have reported earlier on the procedures for magnification fluoroscopy [116], but as the need for this procedure has been found to be rare clinically, we now think it advisable not to apply this technique as frequently as we previously recommended. The fluoroscopic time should be kept as brief as much as possible.

Magnification fluoroscopy can be used in magnification cineradiography.

# *Chapter III*. Clinical Significance of Magnification Radiography

Normal roentgenography is conducted as a routine procedure, but it does not always reveal the details of a lesion, especially the minute findings. The adoption of magnification radiography in addition to normal roentgenography was therefore considered so as to ascertain whether additional findings could be obtained. Various diseases were chosen for study. The patients were placed in the same position as for normal roentgenography and the central X-rays were directed along the same line to the part to be examined. The findings of normal roentgenography and magnification radiography are shown in pairs to permit comparative study. This procedure will make it clear why a small focal-spot tube is used in magnification radiography.

Magnification radiography can now be applied to every part of the body. First, we show cases where magnification radiography has given good results in diagnosing diseases of the soft tissue, lung or bone. Some parts of the body not imaged on the normal roentgenogram can be visualized when contrast medium is used.

Magnification radiographs of the digestive tract where barium sulphate was used are also shown, also bronchi and the lymphatic system visualized by oily contrast media. These are compared with normal roentgenograms taken under the same exposure conditions. Finally, we show angiograms.

In all cases the history and current symptoms of the patients are briefly described and the findings obtained by normal roentgenography are compared with the characteristic findings of magnification radiography.

## A. Magnification Radiography Conducted with Small Focal-Spot Tubes

The image of the normal roentgenogram is essentially poorer in information than that of a magnification radiograph of the same part of the body taken with the appropriate technique.

The normal roentgenogram may appear very sharp (Fig. 40 A, B), yet when enlarged optically it presents images of so rough a tissue structure that they provide far less information than can be obtained from a magnification radiograph.

Note that, even when the roentgenogram is taken normally the combination of screens, the findings are much less than with optical magnification radiography. The thin bone with fine trabeculation is inspected on a normal roentgenogram, taken on non-screened film. This technique is intermediate between magnification radiography and microradiography (Nixon [56]; Rupp and Radke [66]). The optically enlarged 6× magnified radiograph of the bone, where

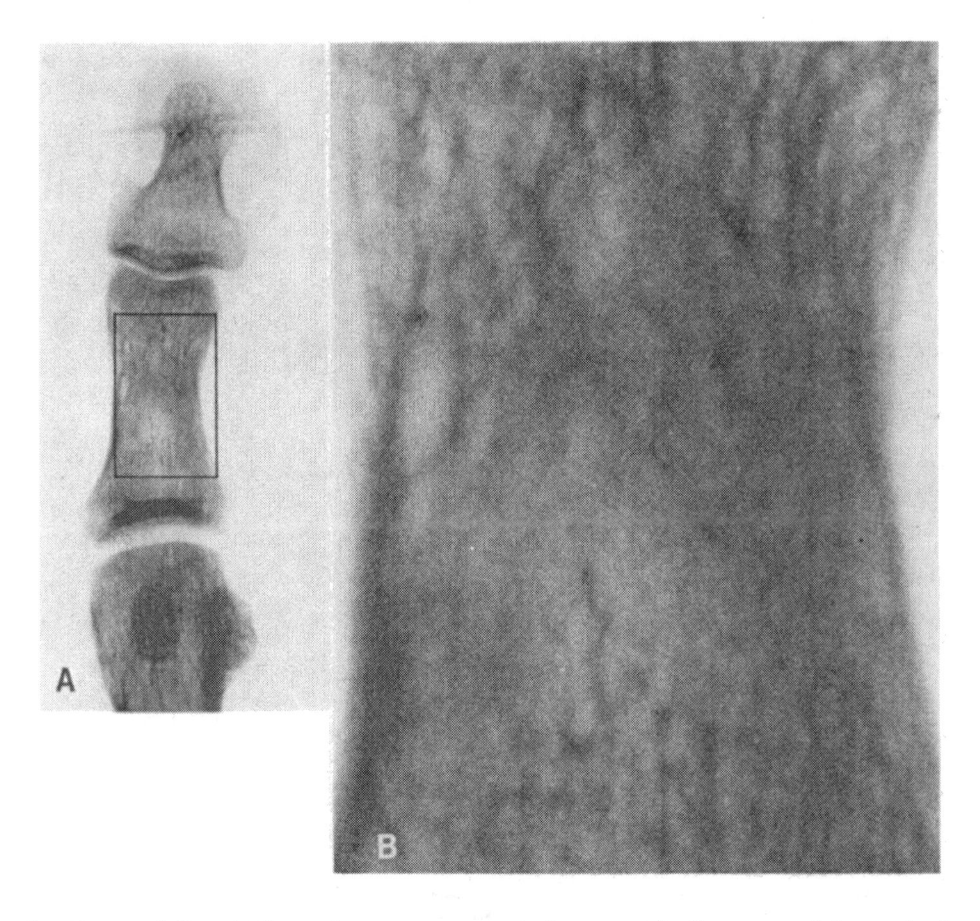


Fig. 40A and B. A) Normal roentgenogram of toe surgically removed from the body. Exposure conditions: 2 mm focal-spot tube, 50 kV, fine-grain screens (FS) and film combination, focal spot-object distance 200 cm, object-film distance zero (contact). B) Optical magnification of Fig. 40A. Although the image quality of the normal roentgenogram looks excellent, in actual fact it is far inferior to that of the magnification radiogram as shown in Figs. 41 B and 44 D

medical X-ray film is used, presents a rough structure of bone trabeculation and rich grain formation on the film (Fig. 41 A, B). High-magnification radiography provides a sharper image than normal roentgenography, even when conducted without intensifying screens (Fig. 44 D).

When 2 × magnification radiography of the same bone is done with X-ray tubes having a focal spot of 2 mm, 0.3 mm, 0.1 mm, and 0.05 mm and the results are compared, the image obtained with the 2 mm focal-spot tube is found to be least sharp (Fig. 42A) and that with the 0.3 mm focal-spot tube is still poor (Fig. 42B). The images obtained with 0.1 mm and 0.05 mm focal-spot tubes are comparable in quality, but the latter (Fig. 42D) is slightly better than the former (Fig. 42C).

The 4 × magnification radiograph of the bone taken with a 0.3 mm focal-spot tube is poor, showing lack of sharpness (Fig. 43 B); that taken with a 0.1 mm focal-spot tube is excellent (Fig. 43C). Nevertheless, when compared with that obtained with a 0.05 mm focal-spot tube (Fig. 43D), the image taken

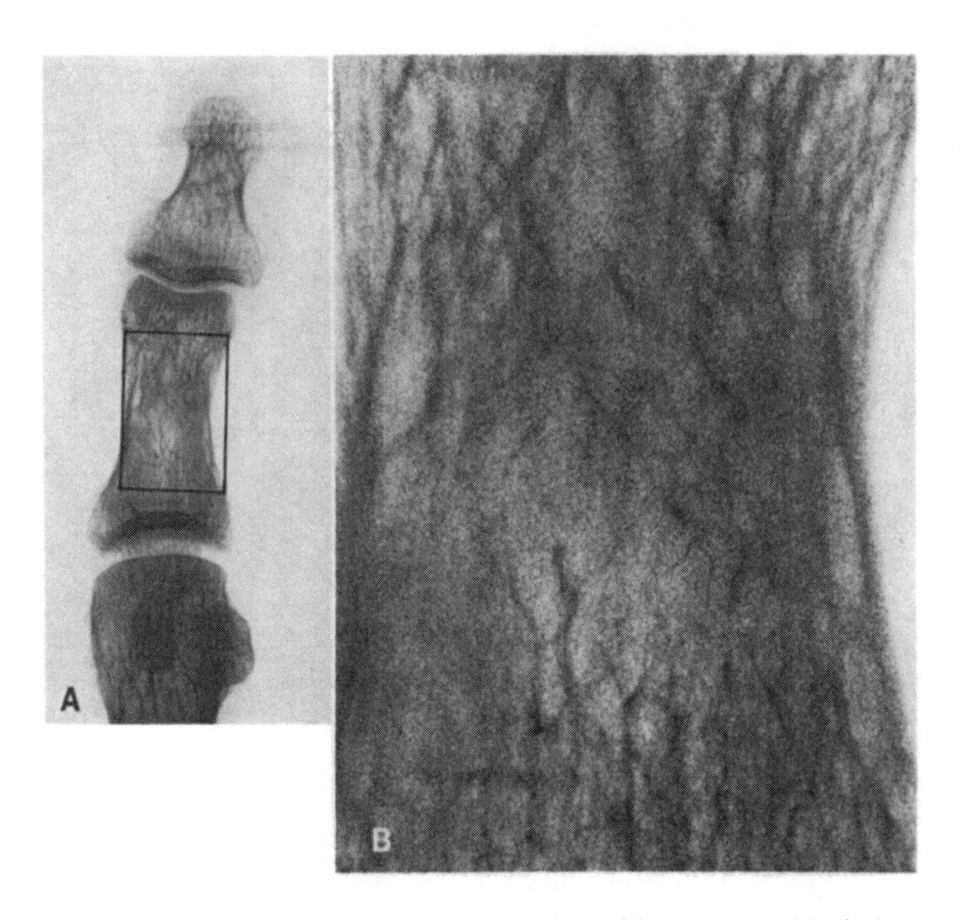


Fig. 41 A and B. A) Normal roentgenogram taken without screens. Medical X-ray film (Fuji KX) was used. Other exposure factors the same as for Fig. 40 A. B) Optical magnification of Fig. 41 A. Image quality is excellent but grain of film interferes with the examination. Image quality is better than in Fig. 40 B, but not as sharp as in Fig. 44 D

with a 0.1 mm focal-spot tube is less sharp (Fig. 43 C).

The 6 × magnification radiograph taken with a 0.3 mm focal-spot tube is very poor in sharpness of image and is useless in practice (Fig. 44 B), while that taken with a 0.05 mm focal-spot tube is good (Fig. 44 D) and the bony trabeculations are clearly seen. When the image obtained with a 0.1 mm focal-spot is examined (Fig. 44 C), it lacks sharpness and is of poorer quality than that taken with a 0.05 mm focal-spot tube.

In short, in magnification radiography of a non-moving organ, such as bone, the 0.05 mm focal-spot tube is superior to the 0.1 mm focal spot tube in providing a sharp image.

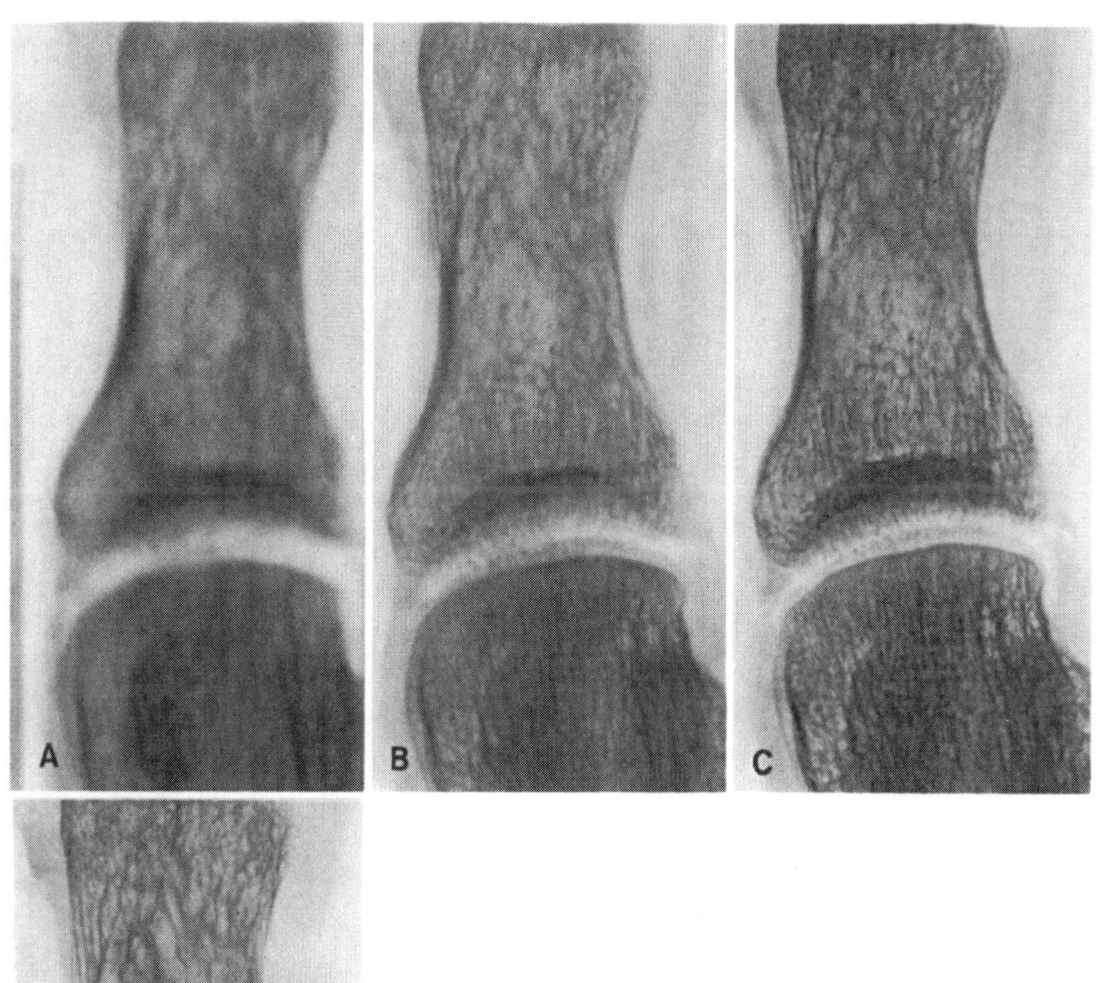


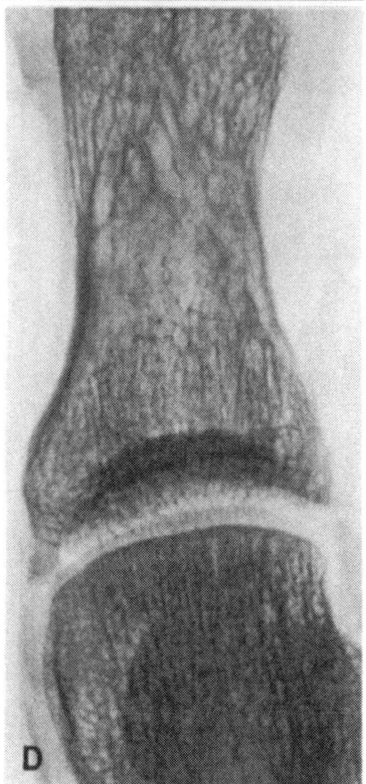


Fig. 42 A–D. 2 × magnification radiograph of toe. A) Taken with 2 mm focal-spot tube. Sharpness of image is very poor. B) Taken with 0.3 mm focal-spot tube. Sharpness of image is bad. C) Taken with 0.1 mm focal-spot tube. Sharpness of image is good. D) Taken with 0.05 mm focal-spot tube. Sharpness of image is excellent

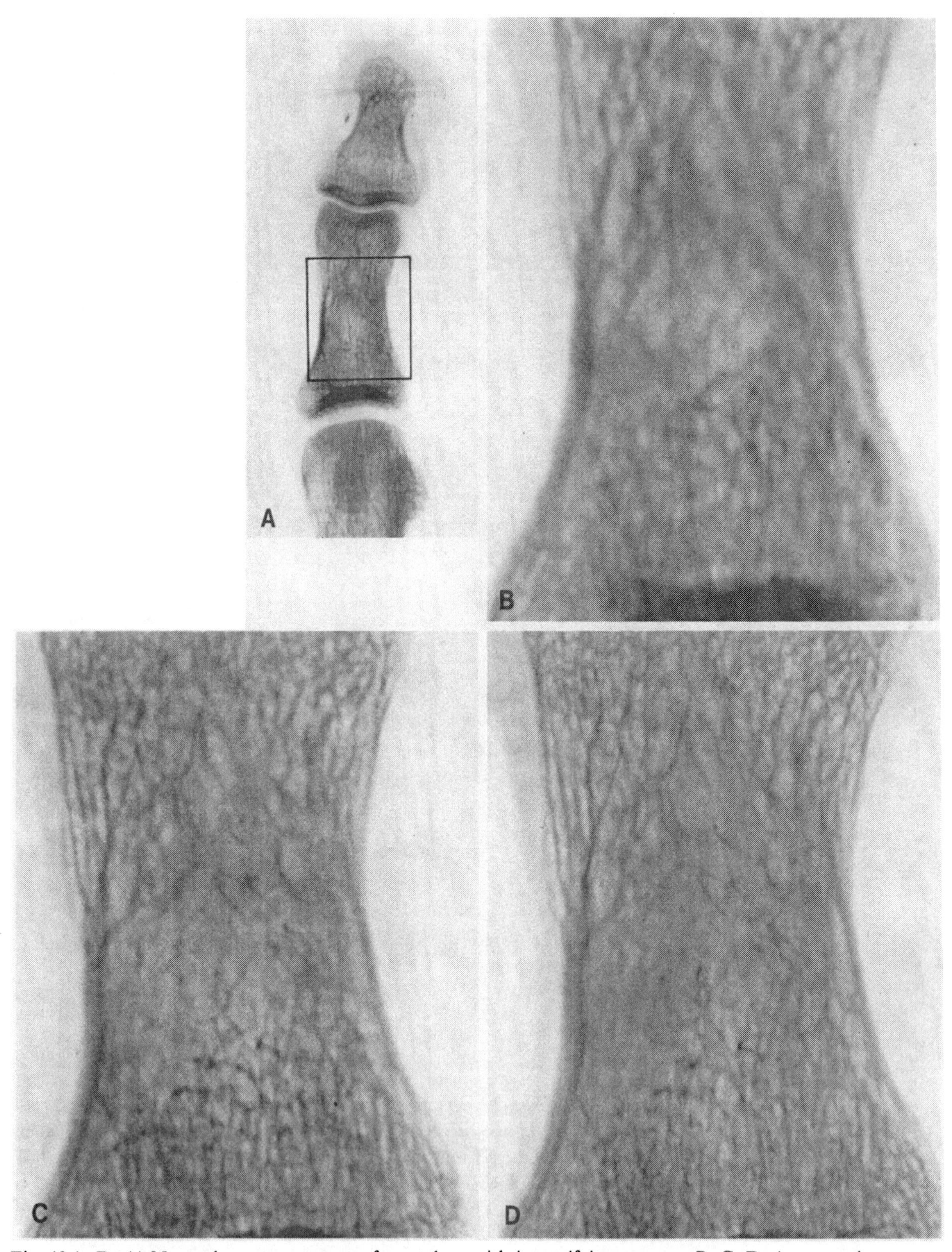


Fig. 43 A–D. A) Normal roentgenogram of toe taken with intensifying screens. B, C, D. 4 × magnification radiographs of toe taken at 80 kVp. B) Taken with 0.3 mm focal-spot tube. Obvious lack of sharpness. C) Taken with 0.1 mm focal-spot tube. Sharpness of image fairly good but inferior to Fig. 43 D. D) Taken with 0.05 mm focal-spot tube. Sharpness of image excellent

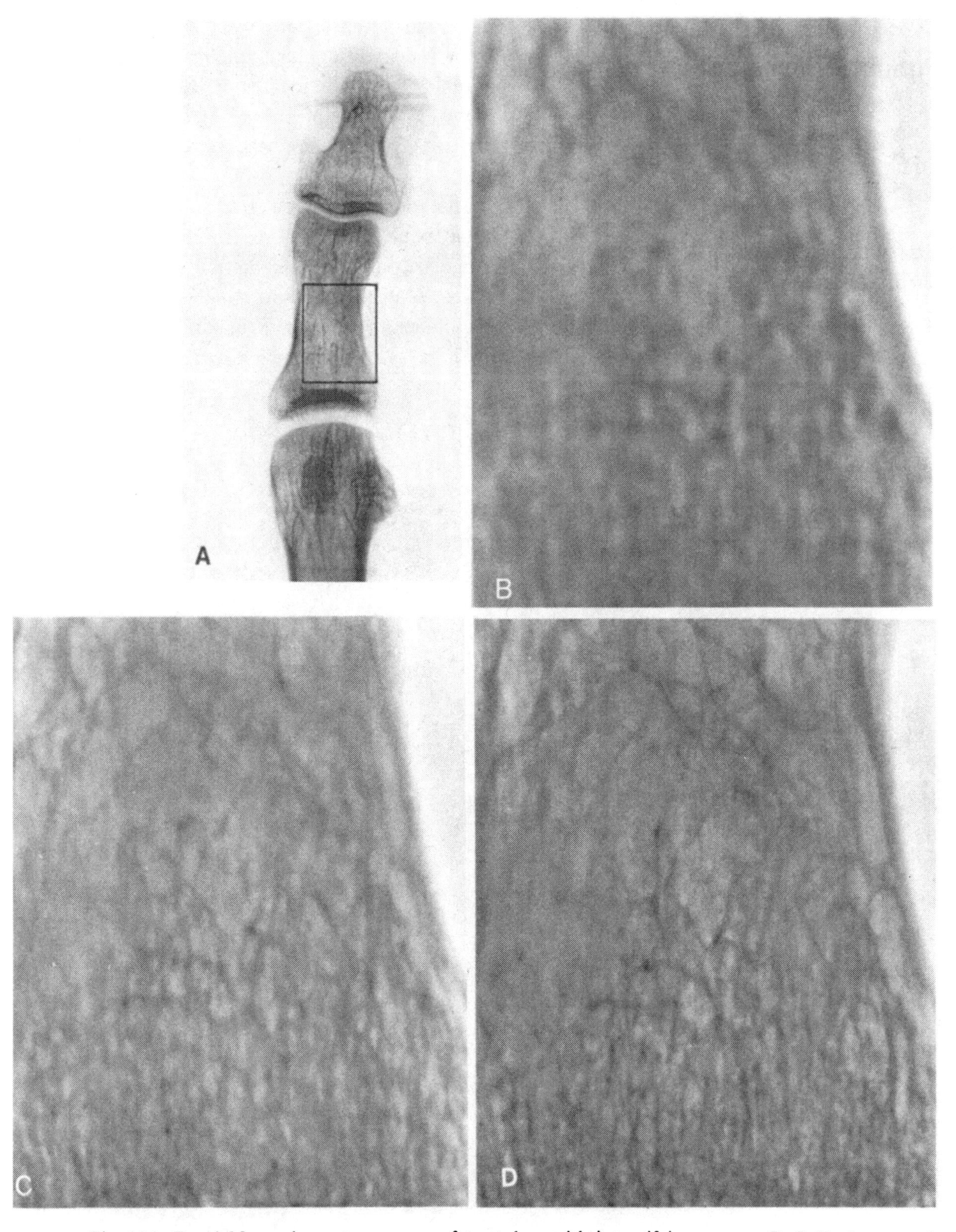


Fig. 44 A–D. A) Normal roentgenogram of toe taken with intensifying screens. B, C, D. 6 × magnification radiographs of toe. B) Taken with 0.3 mm focal-spot tube. Obvious lack of sharpness. C) Taken with 0.1 mm focal-spot tube. Sharpness of image better than Fig. 44 B, but worse than Fig. 44 D. D) Taken with 0.05 mm focal-spot tube. Sharpness of image excellent

## B. Magnification Radiography without Administration of Contrast Media

### Case 1 (Bone): Gout

Male, 67 yrs (Figs. 45 A, B and 46 A, B).

*History:* Main complaint: occasional pains in both big toes.

Normal roentgenography and magnification radiography of the metatarsal bones of the right and left first toes were carried out. Exposure conditions for magnification radiography: 0.05 mm focal-spot tube, FFD 93 cm, 6× magnification, 80 kV, 3 mA, 0.01 sec. Toe thickness was 3 cm.

The normal roentgenogram of the metatarsal bone of the left toe reveals a small irregularity in the joint margin (Fig. 45 A). There is no punched-out shadow. On the magnification radiograph, in addition to the findings of the normal roentgenograms,

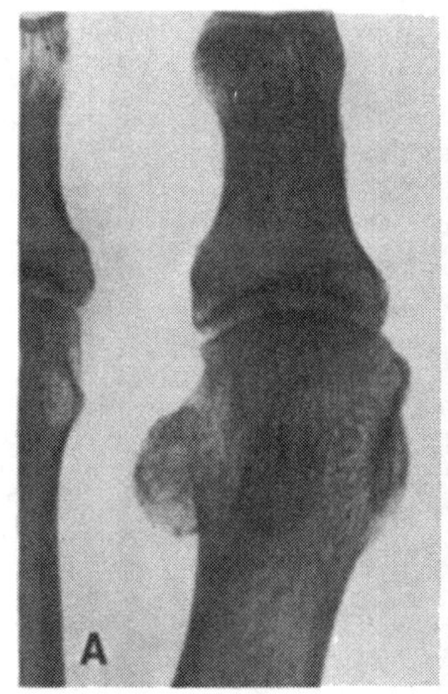


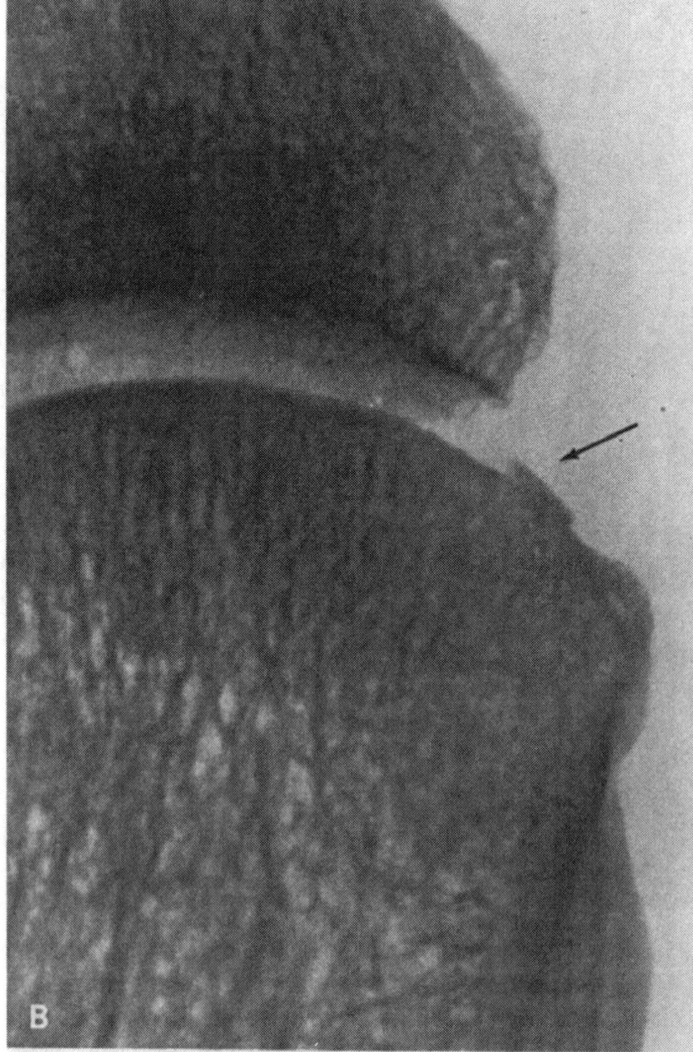


Fig. 45 A. Normal roentgenogram

Fig. 45 B. 6× magnification radiograph

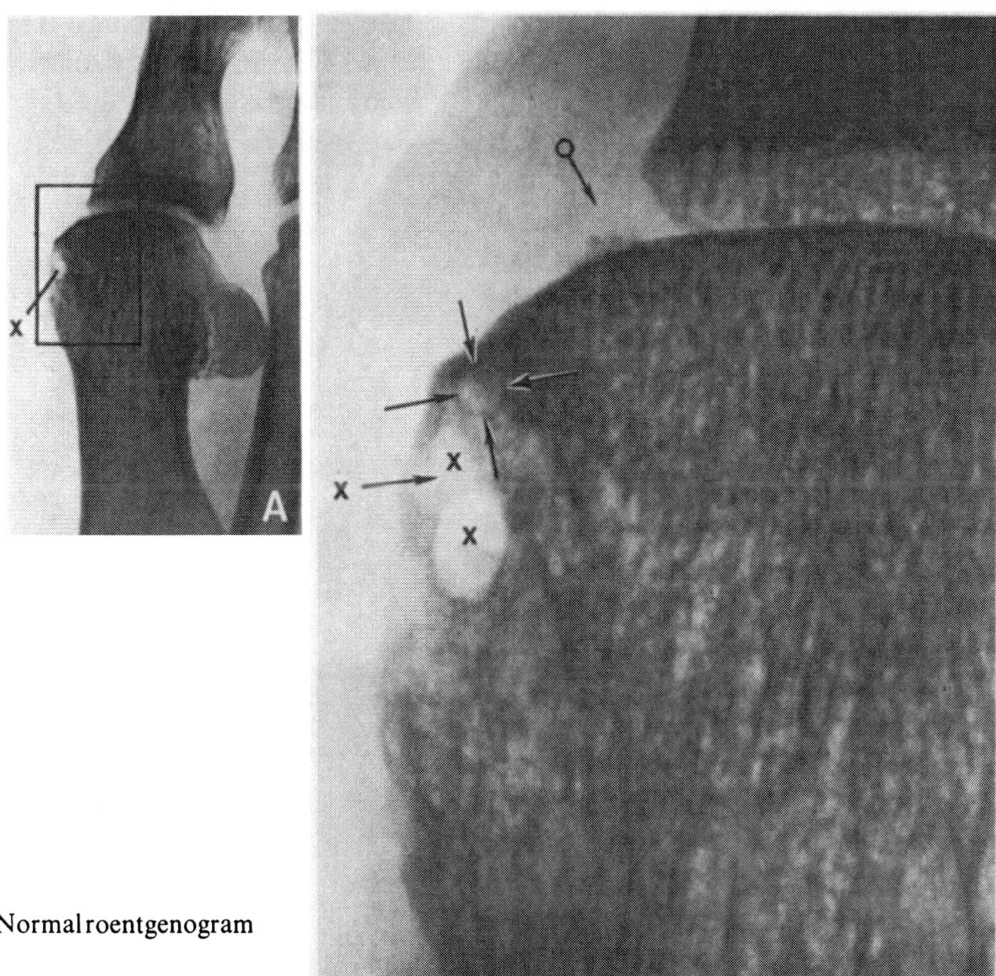


Fig. 46A. Normal roentgenogram

Fig. 46B. 6 × magnification radiograph

there is seen marked deformity and bulging of the joint margin (Fig. 45 B). The calcification of the bone does not appear on the normal roentgenogram but is imaged by magnification radiography (↗).

The normal roentgenogram of the metatarsal bone of the right toe reveals a punched-out lesion of the bone (×) (Fig. 46A). Neither additional punched-out shadow in the bone nor any irregularity of the contour of the bone or joint is seen.

The magnification radiograph reveals three more small punched-out lesions, ×, × and →↓↑←, located just above the lesion (×) seen on normal roentgenogram. There is also in the contour of the joint an irregular small zigzag shadow of calcification (♂) (Fig. 46 B).

## Case 2 (Chest): Silicosis

Male, 49 yrs, plottery worker (Fig. 47 A, B).

*History:* Employed in a plottery since the age of 25. The environment of the work place was very dusty. No history of disease worth mentioning. The main complaint is frequent expectoration but practically no cough.

Normal roentgenography and magnification radiography of the chest were performed. Exposure conditions for the latter: 0.05 mm focal-spot tube, FFD 106 cm, maximal magnification ratio 4 ×, 125 kV, 3 mA, 0.05 sec. Chest thickness 18 cm.

On the normal roentgenogram the pulmonary markings are traceable throughout the entire lung field although a slight increase is recognized in nodular shadows of diameter 0.5 mm to 2 mm (Fig. 47 A). There is slight thickening and lifting of the interlobar pleura. According to the ILO V/C 1971 international classification, the condition is as follows and corresponds to stage-I silicosis.

In the magnification radiograph the small round shadows are scarcely visible, but instead many small, densely distributed ring shadows (♂) of 1 to 3 mm diameter are clearly seen, also small linear shadows (⇗) 0.5 mm wide sometimes running parallel at intervals of 1 to 2 mm and forming fine networks. Interlobar pleuritic scar (↗) and test markers (↗) are seen (Fig. 47 B).

This picture of diffuse fibrosis is apparent over a wide area of the lung field; the finding is interpreted as belonging to the category of a more advanced stage. Thus, most of the small nodular shadows seen by normal roentgenography are found to be overlapping images of fibrosis. The condition should therefore be classified as stage-II rather than stage-I silicosis, and the significance of this contribution is acknowledged.

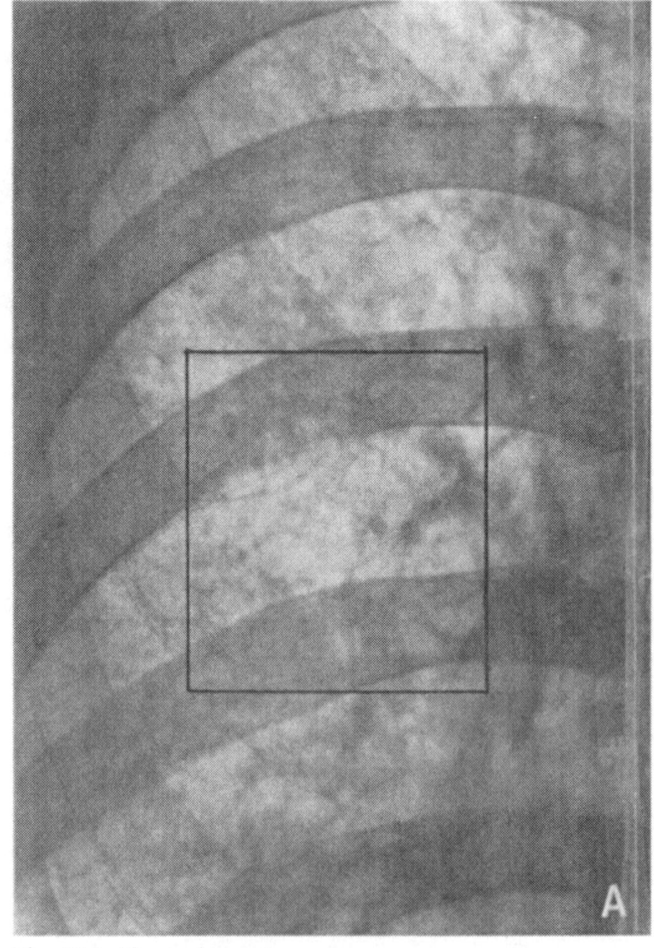


Fig. 47 A. Normal roentgenogram

Fig. 47 B. 4 × magnification radiograph ▷

| Small opacities | | | | | | | | | Large opacities | |
|---|---|---|---|---|---|---|---|---|---|---|
| rounded | | | | irregular | | | | combined | | |
| type | profusion | zones | | type | profusion | zones | | profusion | type | size |
| | | × | × | | | × | | | | |
| P | 1/1 | × | × | S | 1/1 | | | | | 0 |
| | | × | × | | | | × | | | |

| Pleural thickening | | | | Ill-defined diaphragm | Ill-defined cardiac outline | Pleural calcification | | | | Symbols |
|---|---|---|---|---|---|---|---|---|---|---|
| Costo-phrenic angle | site | width | extent | | | diaph | wall | other | grade | |
| 0 | R | a | 1 | 0 | 0 | | | | 0 | 0 |

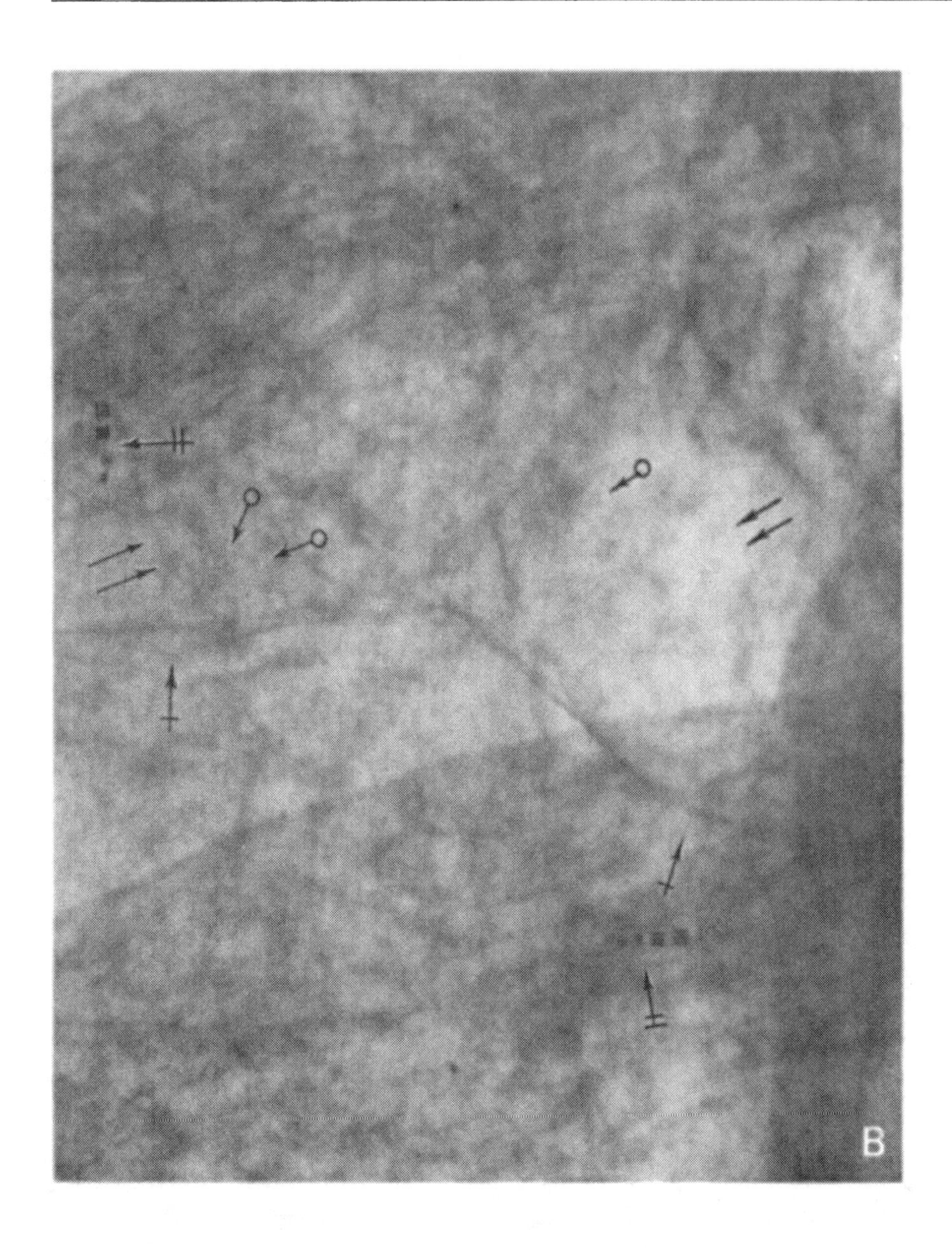

## C. Magnification Radiography with Administration of Contrast Media

### Case 3 (Abdomen): Atrophic Hyperplastic Gastritis and Gastric Ulcer

Female, 67 yrs, housewife (Fig. 48 A, B).

*History:* Epigastralgia with hematemesis of about 200 ml 3 days previously.

*Diagnosis:* Gastric ulcer based on X-ray and endoscopic examinations of the stomach.

A hypotonic double-contrast gastroduodenography was conducted by normal roentgenography and magnification radiography. The exposure conditions of 3× magnification radiography were with 0.1 mm focal spot tube, FFD 120 cm, 110 kV, 15 mA, 0.08 sec. The thickness of the part of the abdomen examined was 16 cm.

The normal roentgenogram reveals a stellate pattern with a fairly deep niche in the posterior wall (♂) of the body (Fig. 48 A). The wall of the niche is clearly seen but the folds are interrupted. The findings are suspected as being malignancy. Along the minor curvature of the stomach what are believed to be gastric areas (↖↑↗) more than 3 mm in diameter are indistinctly seen, but numerous.

Compared with normal roentgenography, the magnification radiographic findings show that the wall (↔) of the niche (♂) is

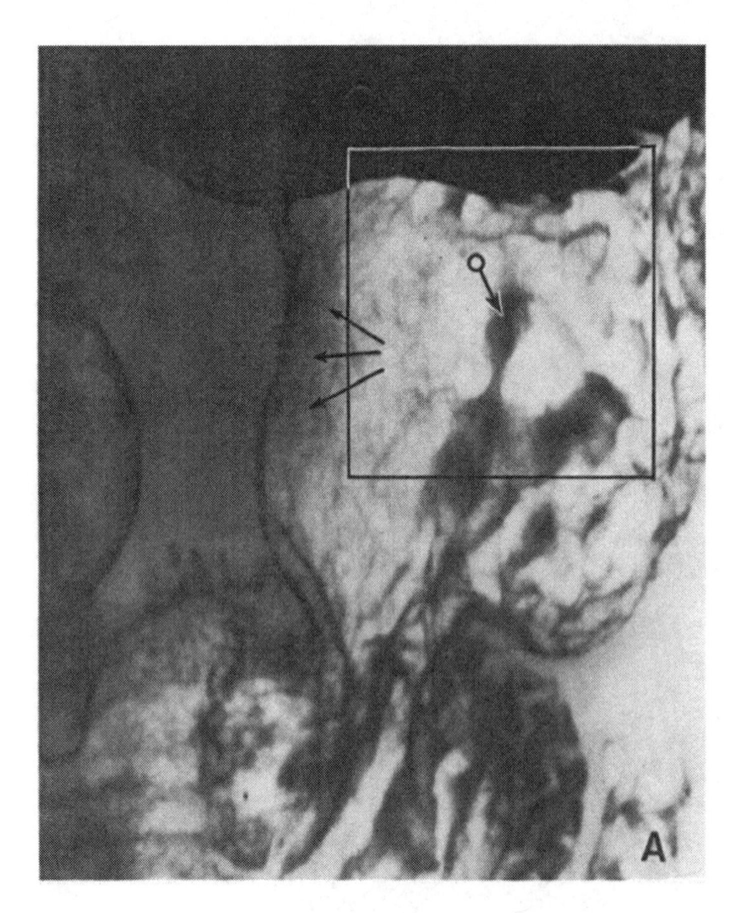


Fig. 48 A. Normal roentgenogram

Fig. 48 B. 3× magnification radiograph ▷

not characterized by changes such as thickened folds. The findings of small changes around the niche and area gastrica are clearly imaged. The fold in the mucosa is not interrupted (↗↙↗↙↗↙) (Fig. 48 B).

The gastric areas (×), seen even on the wall of the niche, consist of irregular bulges that are oval or cocoon-shaped, large and unequal in size. The findings indicate that the condition is due to edema and probably not malignancy. Compared with the normal roentgenographic image, the existence of atrophic hyperplastic gastritis is confirmed by magnification radiography, as the area gastrica are uneven in shape and numerous. This finding was confirmed by gastrocamera examination. The niche healed rapidly after appropriate treatment.

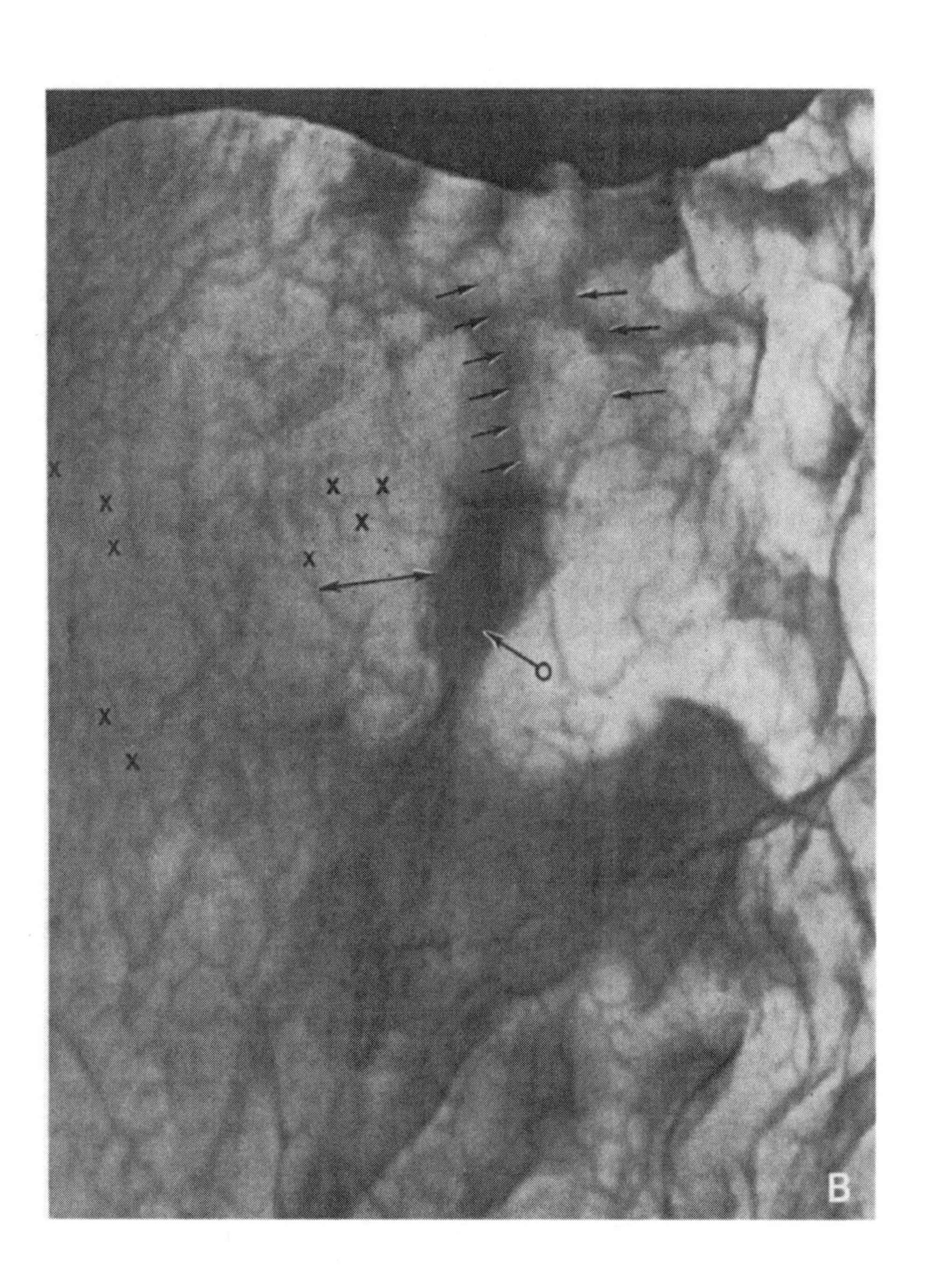

## Case 4 (Chest): Embolism of Oily Contrast Medium in Lung

Female, 56 yrs, housewife (Fig. 49 A, B, C and D).

*History:* No history of illness worth mentioning. Main complaint atypical vaginal bleeding for the previous month.

*Diagnosis*: Carcinoma of uterine cervix.

Lymphography was carried out from both dorsum pedis following KINMONTH's method. Lymphograms show a filling defect of both common iliac nodes.

Normal roentgenography and magnification radiography of the chest were performed 2 hours after lymphography. Exposure factors for magnification radiography: 0.05 mm focal-spot tube, FFD 100 cm, magnification ratio 4×, 120 kV, 3 mA, 0.075 sec. Chest thickness 19 cm.

Normal roentgenography of the chest shows increase of punctiform shadows throughout the entire pulmonary field (Fig. 49 A).

By magnification radiography, both punctiform and hook-like shadows (↑) due to pulmonary embolism are distinctly observed. Twig-like shadows that do not appear on the normal roentgenogram are imaged on the magnification radiograph. The presence of an embolism of oily contrast medium in the peripheral small artery is confirmed. This is due to the resolution effect of magnification radiography (Fig. 49 B).

Lymphography of adult dog from lower extremity with 15 ml of oily contrast media was performed and one hour later the lung was removed. Normal roentgenography was carried out with the exposure conditions 45 kV, 20 mA, 0.05 sec, fine-grain intensifying screens (FS) and FFD 100 cm.

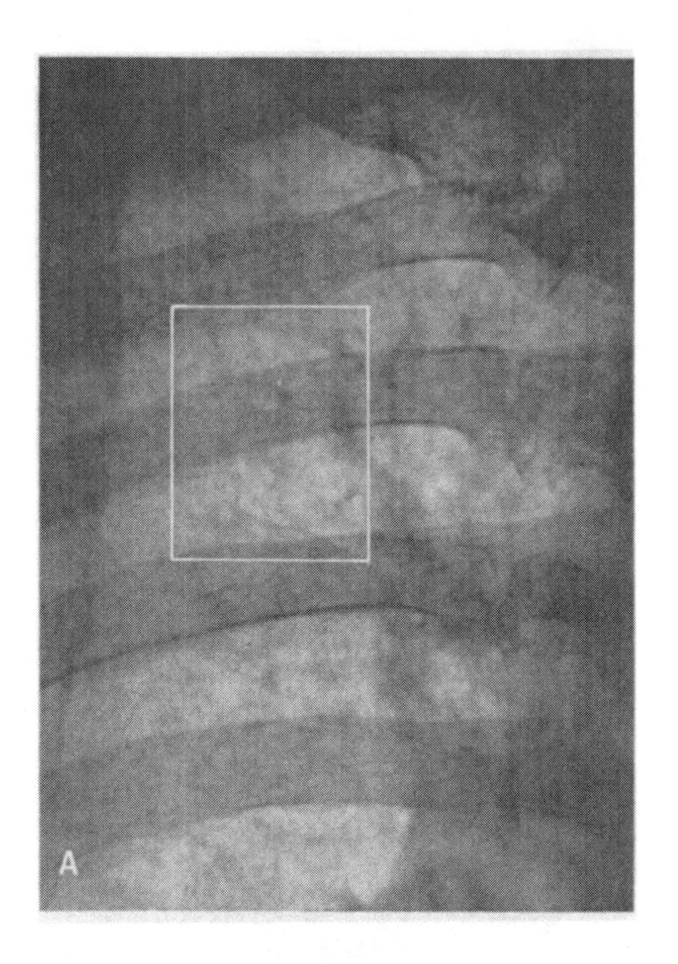


Fig. 49 A. Normal roentgenogram

Fig. 49 B. 4× magnification radiograph ▷

At the peripheral part of the lung artery the hook-like shadows caused by the embolism of oily contrast medium are seen. The findings coincide with those of the magnification radiograph of the patient.

Fig. 49 C shows the normal roentgenogram; 49 D is a 4× optically magnified photo of Fig. 49 C.

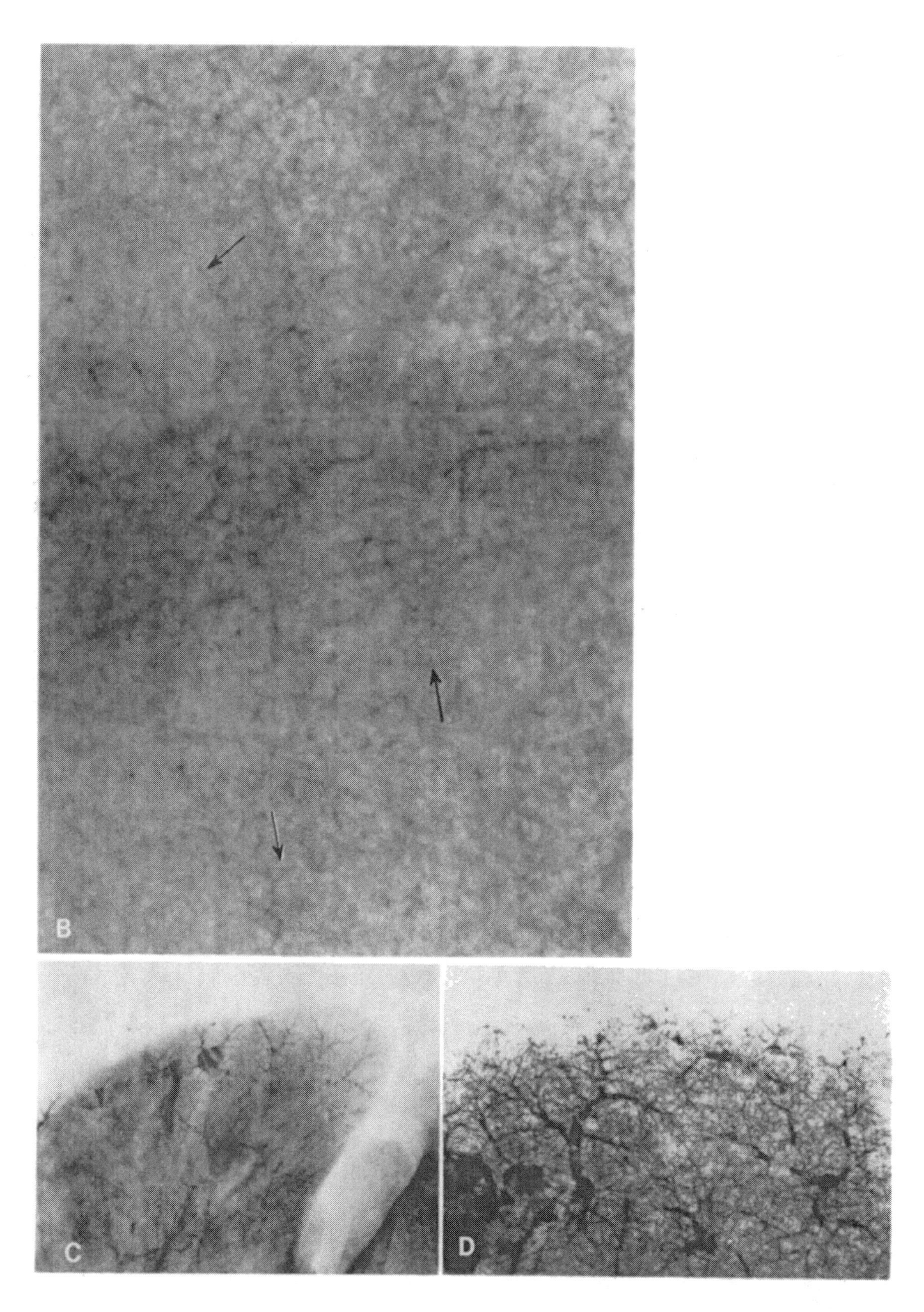


Fig. 49C. Normal roentgenogram of lung of dog

Fig. 49D. 4 × optically magnified photograph of Fig. 49C

## Case 5 (Chest): Bronchial Asthma

Female, 25 yrs, clerk (Fig. 50 A, B, C, D, E and F).

*History:* Innate tendency to catch cold, but no marked abnormality. For last three years asthmatic attacks in spring and autumn. Pulmonary function tests showed both vital capacity and maximal expiratory flow to be normal. The 1-sec vital capacity was 74% and indicated slight obstructive disease. The air-trapping index at 5.4% was somewhat high. After inhalation therapy the 1-sec vital capacity rose to 90% and the air-trapping index became normal.

Normal chest roentgenography shows the pulmonary markings of the lung field to be near normal. Thickening of the wall of the lobar bronchus is seen, and the diaphragm tends to drop with both inspiration and expiration (Fig. 50 A).

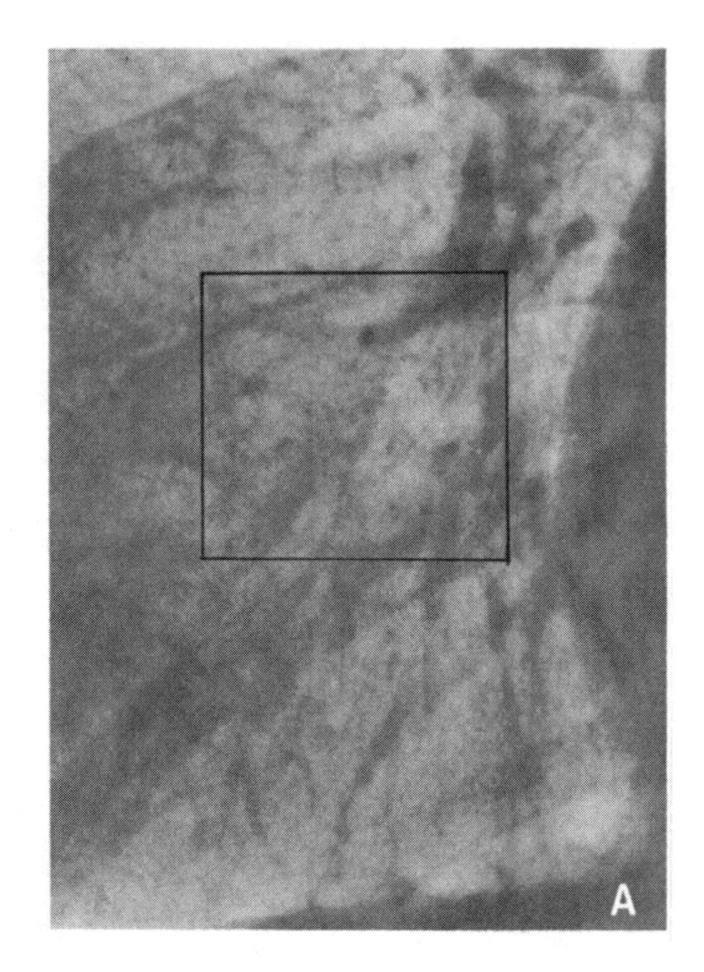

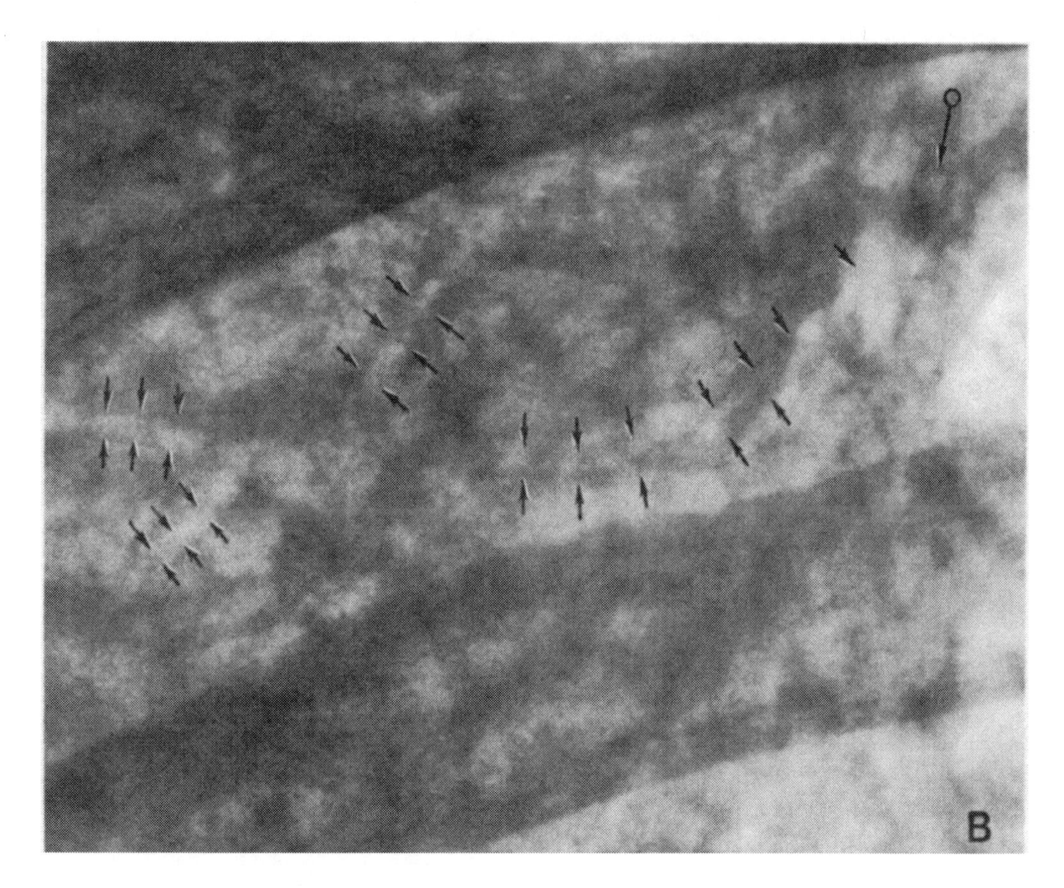


◁ Fig. 50 A. Normal roentgenogram

Fig. 50 B. 4 × magnification radiograph

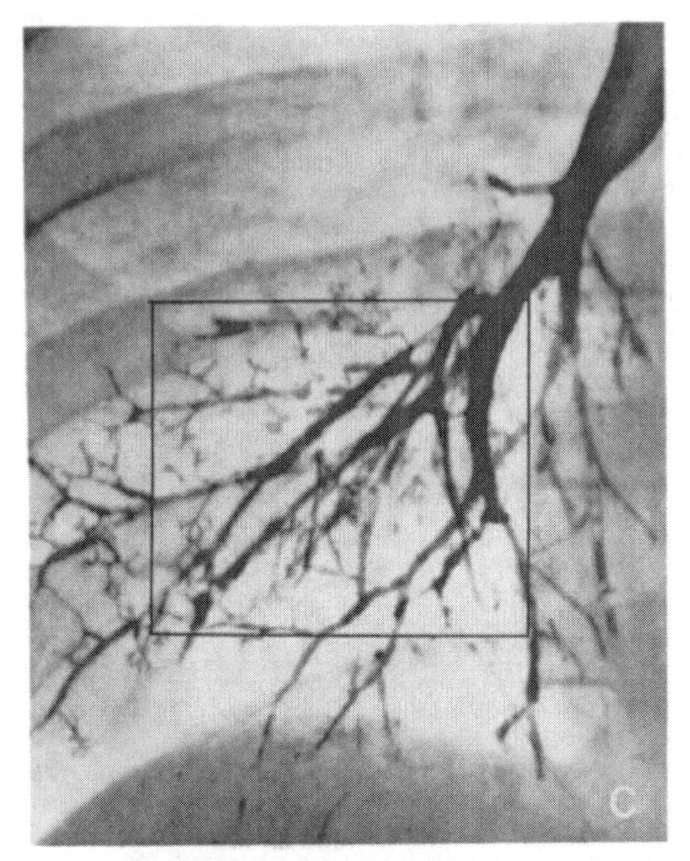

Fig. 50C. Normal bronchogram

Fig. 50D. 4 × magnification bronchogram

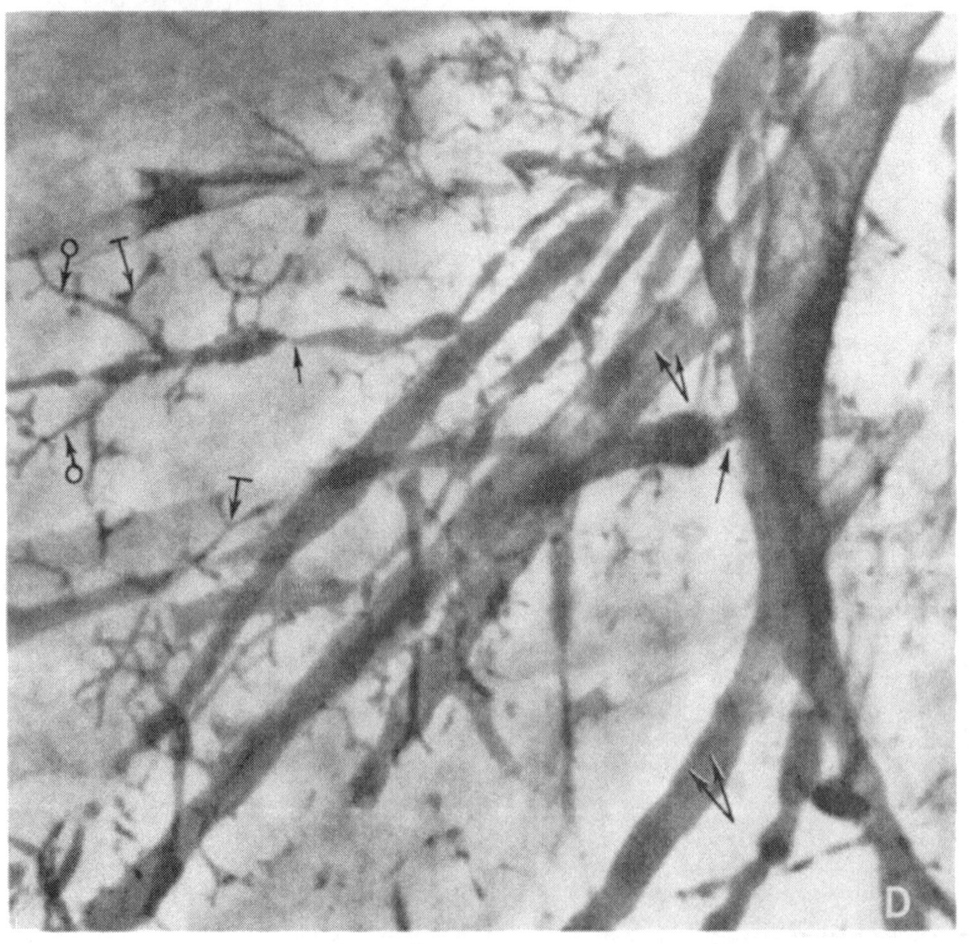

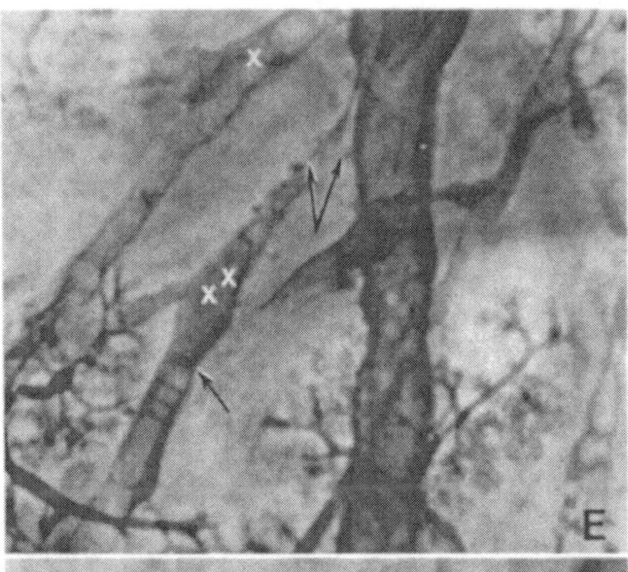


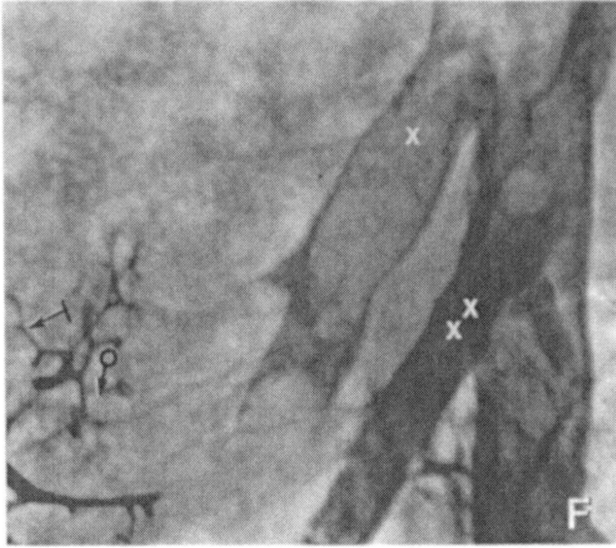


Fig. 50 E. 4 × magnification bronchogram taken before administration of dilator agent

Fig. 50 F. 4 × magnification bronchogram taken after administration of dilator agent

Magnification radiography was conducted with a 0.05 mm focal-spot tube and exposure conditions: 120 kV, 3 mA, 0.05 sec. Chest thickness 17 cm.

On the magnification radiograph a bronchial thickening is seen around the ramus medius lateralis (bronchus in side view ↗↗↗↙↙↙, and bronchus in tangential view ⍴) indicating inflammatory changes (Fig. 50 B).

Bronchography was conducted by normal roentgenography and magnification radiography, the exposure conditions of the latter being: 0.05 mm focal-spot tube, 120 kV, 3 mA, 0.06 sec.

The normal roentgenogram shows an irregularity of the bronchial wall, but the state of the mucosa and the condition of the bronchioles are not clear (Fig. 50 C). Hence, it is not possible to determine whether the irregularity of the wall is due to spasm or to organic changes in the wall.

The magnification radiograph reveals the mucosa to consist of longitudinal mucosal folds (↖ ↗) and the irregularity of the wall to be due to ring-like strangulation (↗). Similar changes are seen in the bronchioles (♂). Also, strangulation is noted at every bifurcation of the bronchi (⤤). The changes are thus considered due to spasm (Fig. 50 D). After inhalation of a bronchial dilator agent the image of the longitudinal mucosal folds disappears, so confirming that the changes are due to spasm. Magnification radiography, however, provides images of spasm (♂, ⤤) extending to the bronchioles (Fig. 50 E and F).

× Anterior basal branch of the lower lobe.

× × Lateral basal branch of the lower lobe.

## Case 6 (Chest): Chronic Obstructive Pulmonary Disease

Male, 69 yrs (Fig. 51 A and B).

*History:* For 4–5 years has had expectoration and dyspnea, also asthmatic attacks in spring and autumn.

Pulmonary function test showed a 1-sec vital capacity of 34% and an expiratory reserve volume of 10%. A diagnosis of chronic obstructive pulmonary disease was established.

Normal roentgenography of the chest shows diffuse linear, dotted and flecked shadows throughout the entire lung field. Pulmonary arterial markings are absent, except in the trunk parts.

A selective bronchography was made by means of a catheter used for selective angiography, and the parts from the bronchioles to the alveolae were radiographed. Exposure conditions for 4 × magnification radiography: FFD 106 cm, 125 kV, 3 mA, 0.05 sec. Chest thickness 18 cm.

Normal roentgenographic findings show the bronchial wall to be irregular (↗) and the alveolae coarse and large (♂) (Fig. 51 A). The state of the walls of the bronchioles could not be determined. A tentative diagnosis would be chronic bronchitis.

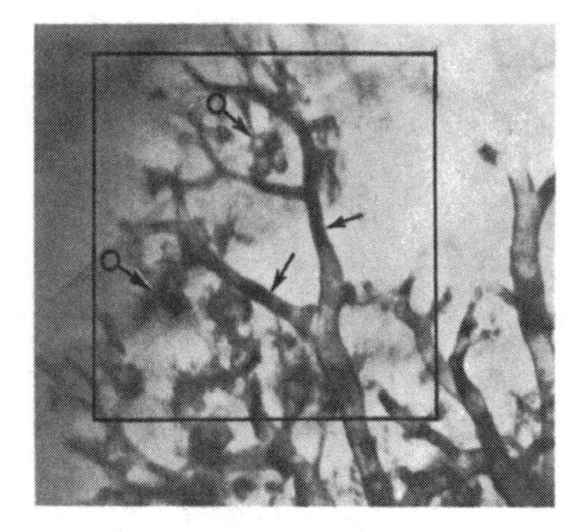

Fig. 51 A. Normal roentgenogram

Fig. 51 B. 4 × magnification radiograph ▷

Magnification radiography shows up the irregularity of the bronchial wall (↓) more clearly, and the change is seen to extend to the walls of the bronchioles (♁) (Fig. 51 B). The wall of the bronchiole is irregular and the part between the bifurcations is narrowed (↥). Here and there in the lobules are dilatations of alveolar sacs (⍏) due to destruction of the alveolar wall.

In addition to the findings of chronic broncho-bronchiolitis, it is clear from the magnification radiographic findings that there is also centrilobular emphysema, diagnosis of which is usually difficult until a pathological examination is performed.

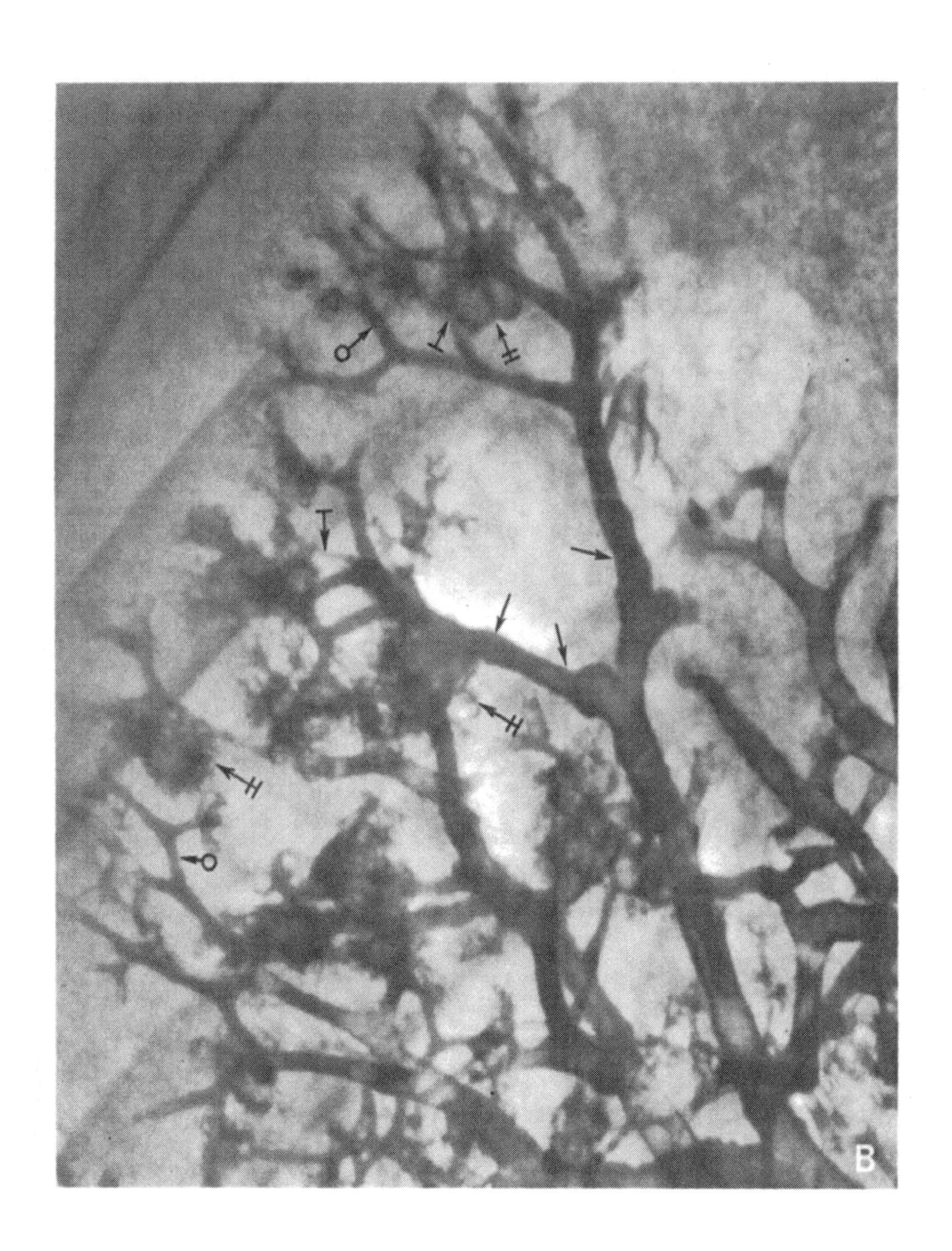
B

## Case 7 (Salivary Gland): Sjögren Syndrome

Female, 36 yrs, housewife (Fig. 52 A and B).

*History:* Swelling of right parotid area, dryness of oral cavity and pain in joints for previous 2 months.

Autoimmune sialosis was suspected. An 18-gauge soft plastic intravenous catheter was inserted into the right stenone duct and Urografin 76%, 1.5 ml, was introduced. Normal and magnification radiography was done. Exposure conditions for magnification radiography: 4× magnification with a 0.05 mm focal-spot tube, FFD 118 cm, 120 kV, 3 mA, 0.05 sec. The thickness of part radiographed was ca. 6 cm.

The normal roentgenogram reveals normal findings, except that the intralobular ductule is stretched and presents the "pruned-tree" appearance (Fig. 52 A).

On the magnification radiograph the intralobular ductules containing contrast medium right to the tips are observed with ease (Fig. 52 B). No irregularity of the wall is noted. At the tip of the narrowed duct are seen small dot-shaped acini (↗) and rather larger ones with irregular contours (♂). The image of the acini is not diffuse and each can be clearly observed. In some parts subalveolar contrast material dissections are seen (✓↗). These are visible on the normal roentgenogram in advanced cases; but not at such an early stage.

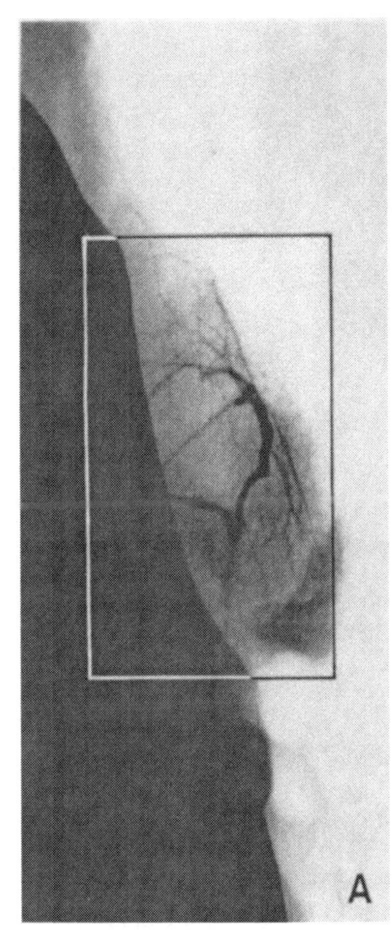


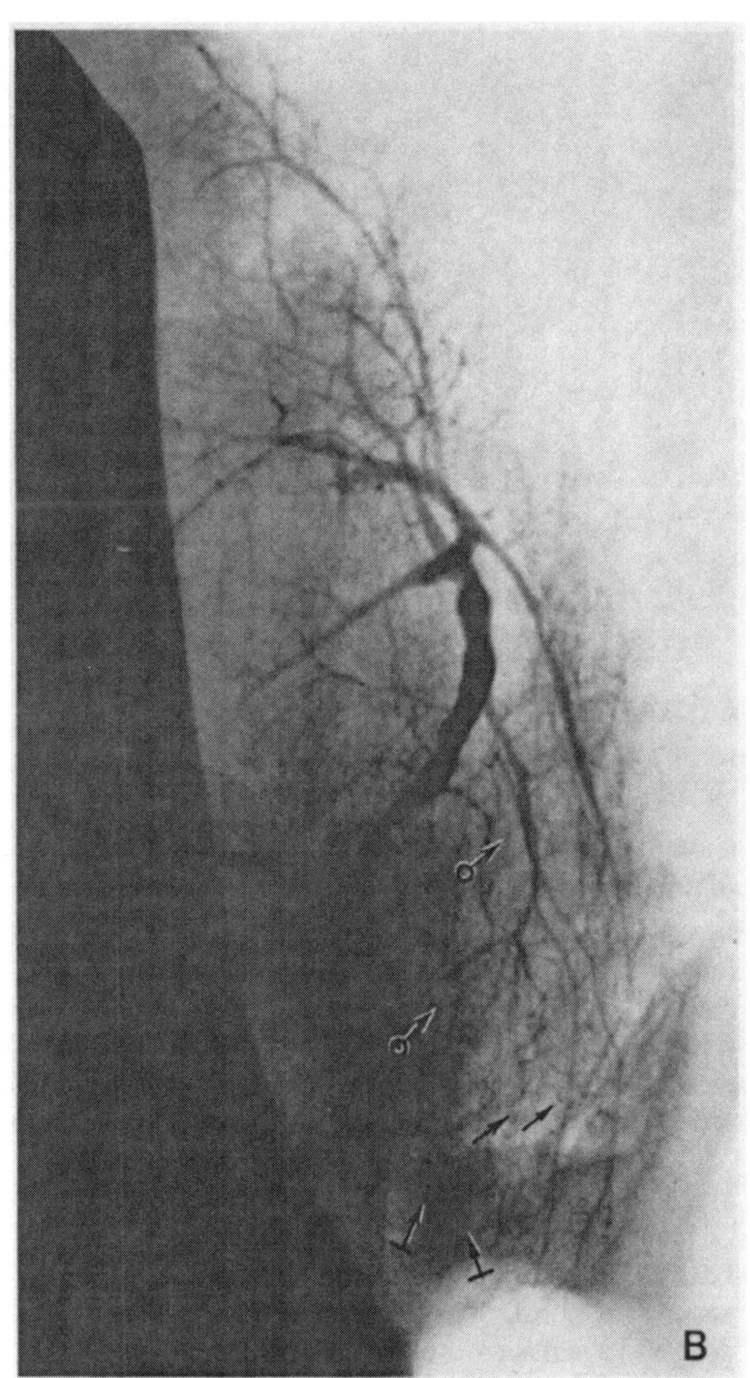


Fig. 52A. Normal roentgenogram

Fig. 52B. 4 × magnification radiograph

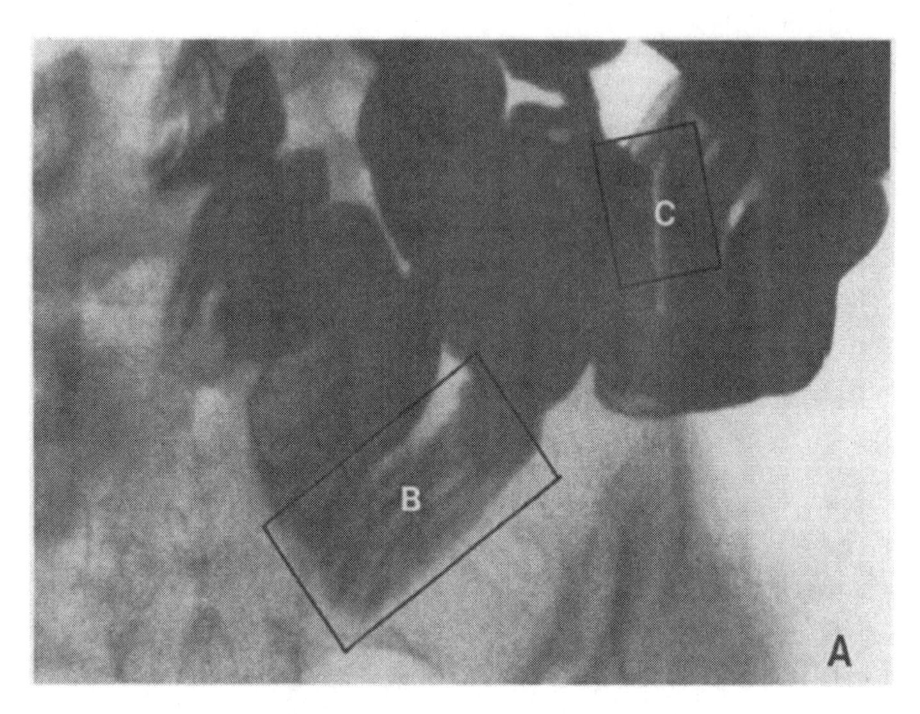


Fig. 53 A. Normal roentgenogram

## Case 8 (Abdomen): Mucosal Pattern of Small Intestine

Female, 18 yrs, housewife (Fig. 53 A, B and C).

*History:* Diarrhoea for 1 month; abdominal pain at times. 250 ml of barium sulfate of 100 w/v % was administered.

Magnification radiography conditions: 0.05 mm focal-spot tube, FFD 100 cm, 120 kV, 3 mA, 0.3 sec. Magnification ratio was 4×. Normal roentgenogram reveals an almost sharp contour of the image of the intestine. Folds are normal. Intestine is pushed upward by tumor (Fig. 53 A).

By magnification radiography (Fig. 53 B) the image of the small intestine is contoured with brush-like shadows (♂); this is the profile view of the villi of the small intestine. The en face view is a small round or clubby image (↗). The diameter of the villi ranges from 0.3 mm to 0.6 mm, the height being from 0.5 mm to 1 mm. Actual size was therefore calculated to be from 0.07 to 0.2 mm in width and 0.1 mm to 0.3 mm in height. The images are considered to be due to the $BaSO_4$ located between the villi. These findings are imaged exclusively on the magnification radiographs. Test marker (↗). Profile image of villi is clearly seen (↗) (Fig. 53 C).

Fig. 53 B. 4× magnification radiograph ▷

Fig. 53 C. 4× magnification radiograph ▷

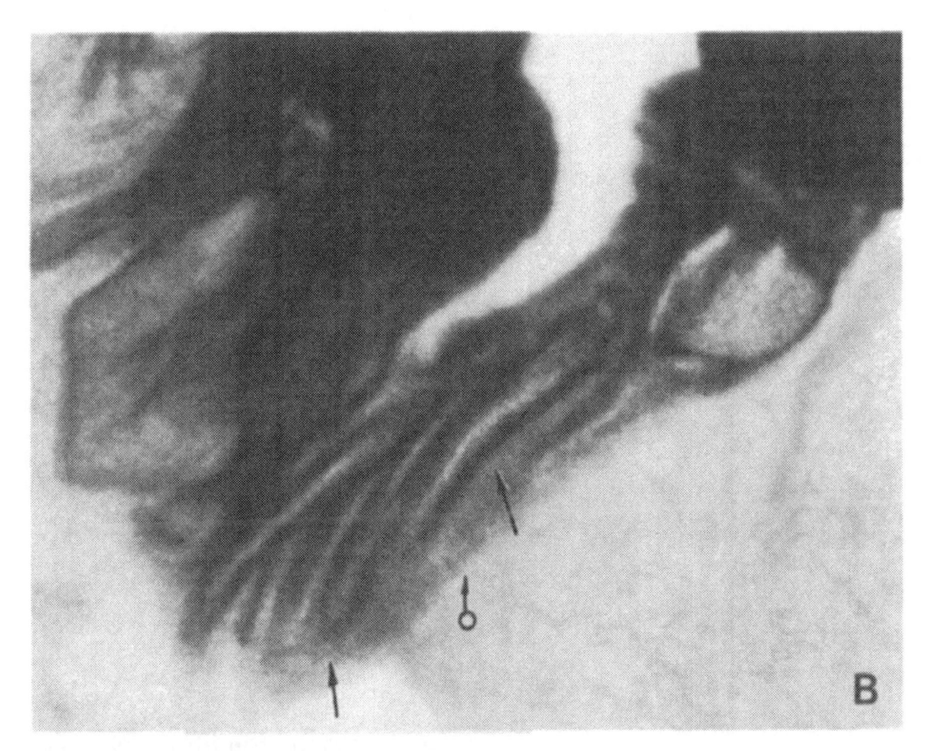
B

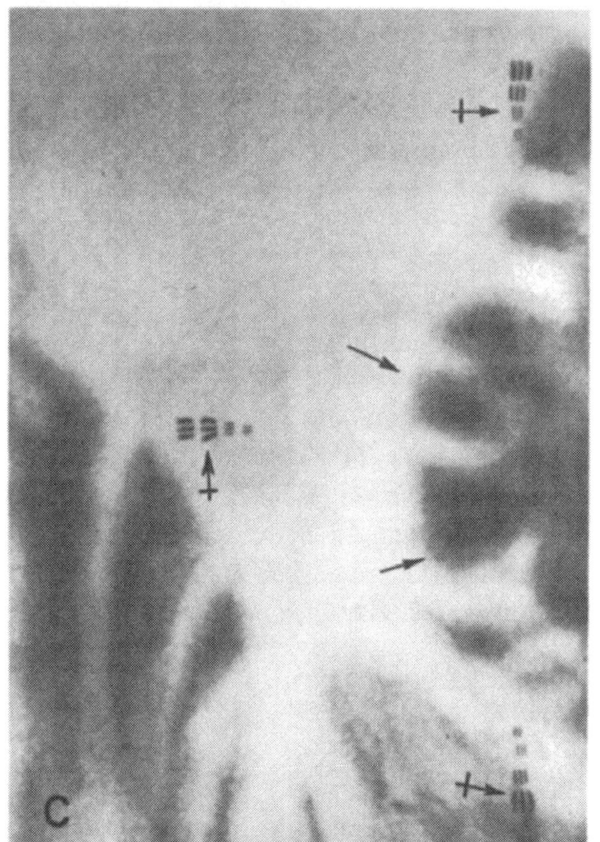
C

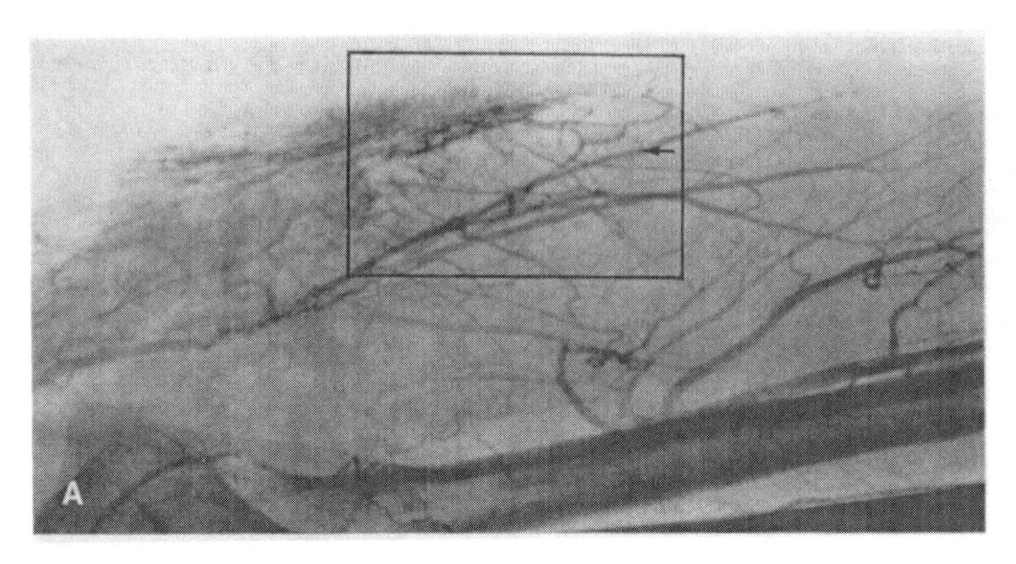


Fig. 54A. Normal angiogram

## D. Magnification Radiography of Vessels

### Case 9 (Popliteal Fossa): Carcinoma of the Skin

Male, 34 yrs (Fig. 54A and B).

*History:* When 5 years old, the patient sustained a scald, resulting in a 15 × 18 cm cicatrix extending downward from the popliteal fossa of the right leg. About 2 months prior to admission an erosion, about 5 × 9 cm in area, appeared at the site. This became nodular and grew rapidly in size.

Femoral arteriography was conducted by normal roentgenography and magnification radiography. Exposure conditions of magnification radiography: 0.05 mm focal-spot tube, 120 kV, 3 mA, 0.1 sec, FFD 106 cm. Leg thickness 10 cm.

The normal angiogram in the arterial phase reveals tortuous vessels of uneven contour in the tumor bed; the blood vessels within the tumor are poorly defined (Fig. 54A).

On 4 × magnification radiograph several branches of vessels within the tumor are seen to radiate into the lobulation. Brush-like shadows of the small blood vessels that are barely visible on the normal roentgenogram are imaged by magnification radiography. The presence of malignancy is confirmed. Test markers (↗) (Fig. 54B).

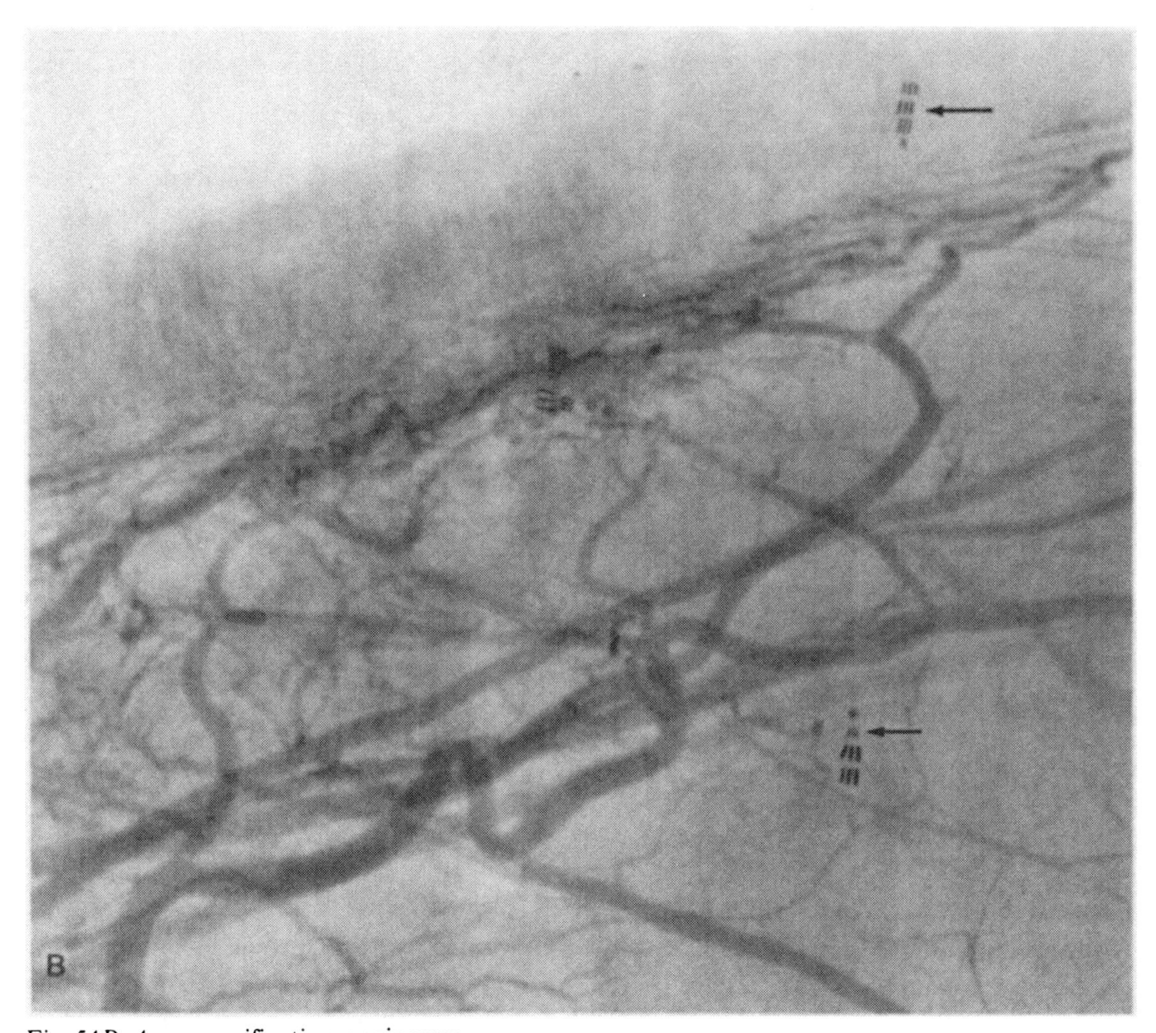

Fig. 54 B. 4 × magnification angiogram

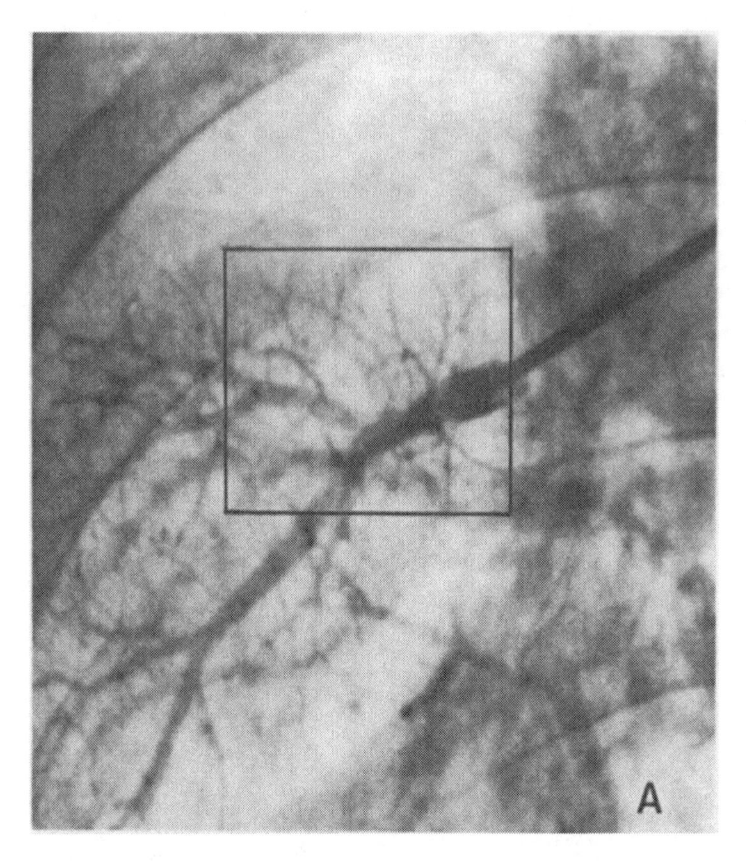


Fig. 55 A. Normal roentgenogram

## Case 10 (Chest): Chronic Obstructive Pulmonary Disease

Male, 64 yrs (Fig. 55 A and B).

*History:* Tendency to catch cold, but no particular history of disease. Dyspnoea during exercise for 3 to 4 years.

Pulmonary function test showed a vital capacity of 1900 ml with a rate per sec of about 34%. The pulmonary scintigram shows the distribution in the middle lobe to be coarse and not homogeneous.

Normal roentgenographic findings of the chest show depression of the diaphragm, indistinctness of the pulmonary markings and increased fibrotic changes.

Selective pulmonary arteriography was performed by normal roentgenography and magnification radiography. Exposure conditions: 0.05 mm focal-spot tube, 120 kV, 3 mA, 0.075 sec. Chest thickness 19 cm.

The normal pulmonary arteriogram shows the diameter of the pulmonary artery and the manner of branching to be normal (Fig. 55 A). There is, however, a slight early return into the pulmonary vein.

The magnification radiograph reveals the normal diameter of the pulmonary artery, but some bending (↗) is seen at each branching point, suggestive of expansion of the alveolae (Fig. 55 B). In the lobular artery (♂) there is an irregularity of the wall and bending that cannot be recognized on normal roentgenogram. The findings indirectly manifested the existence of pulmonary emphysema.

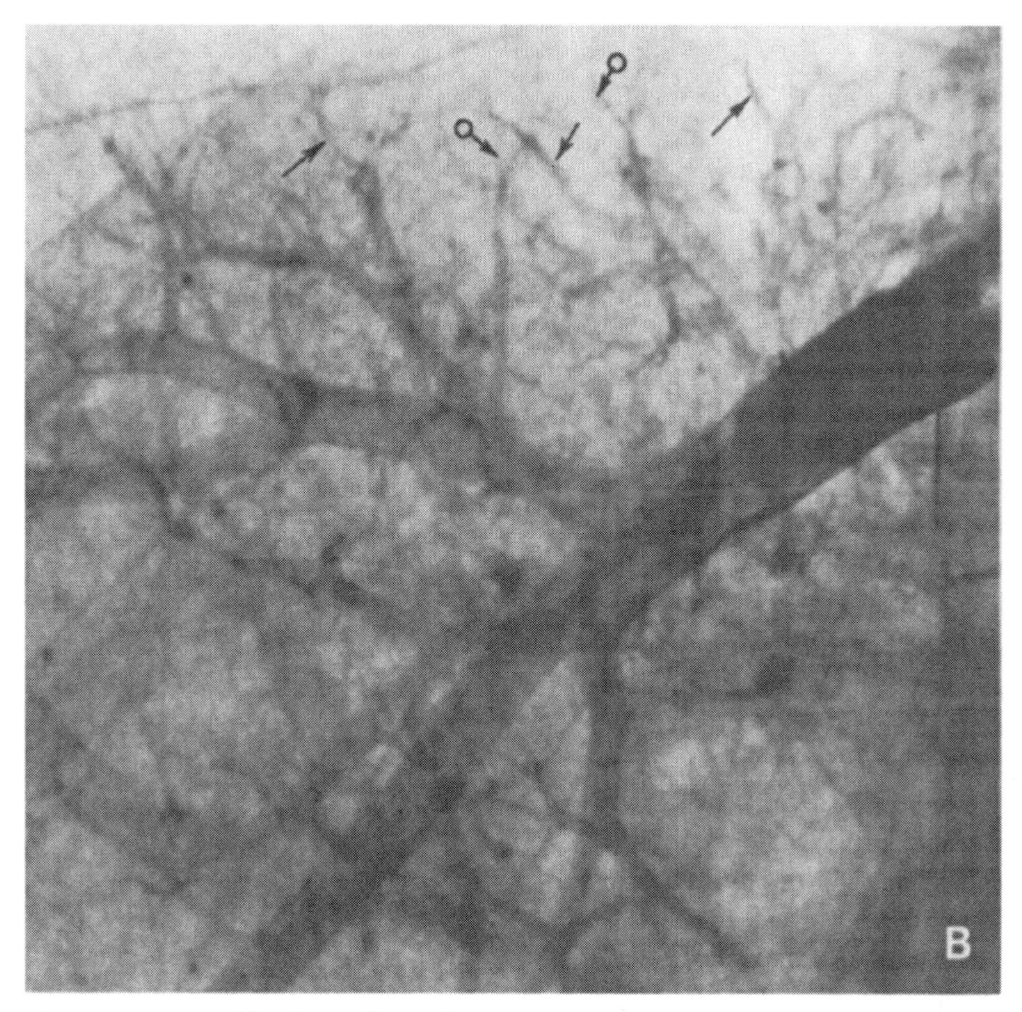


Fig. 55 B. 4 × magnification radiograph

## Case 11 (Chest): Pulmonary Carcinoma

Male, 54 yrs, businessman (Fig. 56 A and B).

*History:* No record of illness worth mentioning.
Main complaints: hemoptysis, cough, and pain in the right chest for 1 month.

The normal roentgenograph of the chest reveals a homogeneous shadow extending outward fan-like from the left hilum, and immediately below this a round, homogeneous shadow of area about 6 × 8 cm. The lower branch of the left pulmonary artery is of low density. The existence of lung cancer was suspected but there was a need to differentiate between inflammation and granuloma.

Selective arteriography of the right bronchial artery was done by normal roentgenography and magnification radiography.

For magnification radiography it is usual to use a 0.05 mm focal-spot tube, but in this case a 0.1 mm focal-spot tube was used. Exposure conditions: FFD 120 cm, magnification ratio 3 ×, 110 kV, 10 mA, 0.08 sec. Five serial radiographs were made at one frame per sec. Chest thickness in the diseased area: 17 cm.

Normal roentgenography reveals along the wedge-shaped homogeneous shadow a large, narrowing artery, although tapering of the blood vessels is not lost. There is no irregularity of the vessel wall, presenting a picture of inflammation (Fig. 56 A).

Magnification radiography reveals, besides the above, interruption of large vessels, hypervascularity in the peripheral part of the tumor shadow, tortuous and narrow image of the vessels invading the tumor shadow, and punctiform and flecked shadows intermingled. Thus, a diagnosis of malignant tumor is established. Hypovascularity is noted at the center of the tumor (Fig. 56 B).

The image of the blood vessels that appeared blurred and could not be confirmed by normal roentgenography is visualized by magnification radiography and contributes to the diagnosis of this particular case.

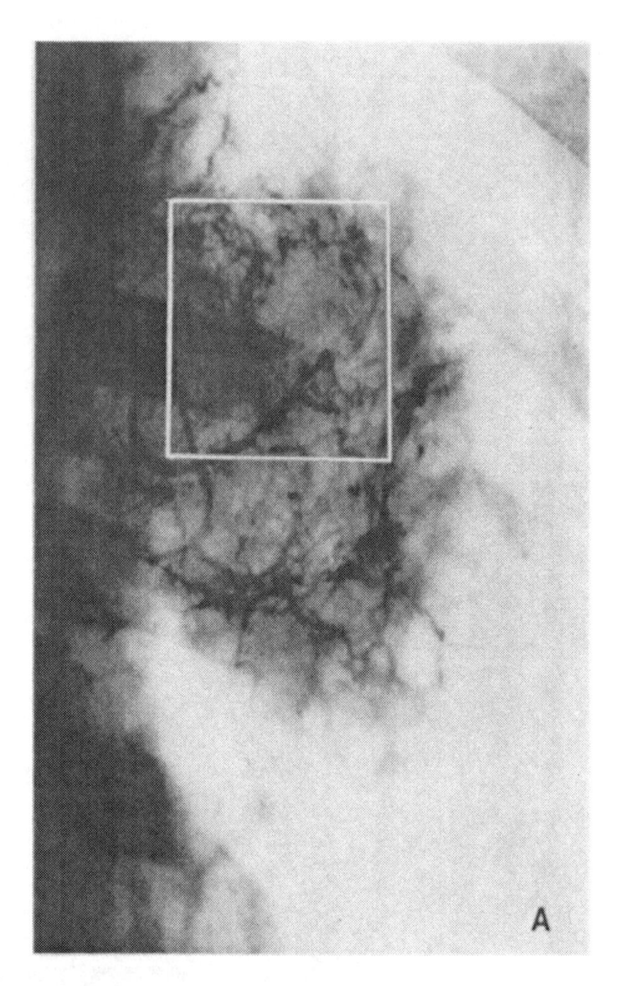


Fig. 56 A. Normal roentgenogram

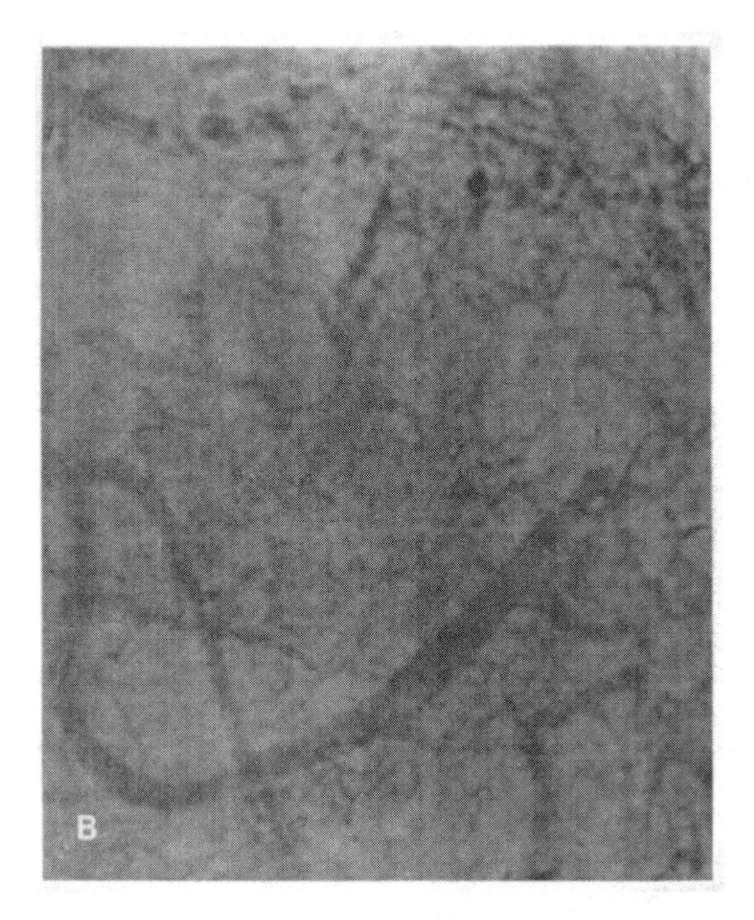


Fig. 56 B. 4 × magnification radiograph

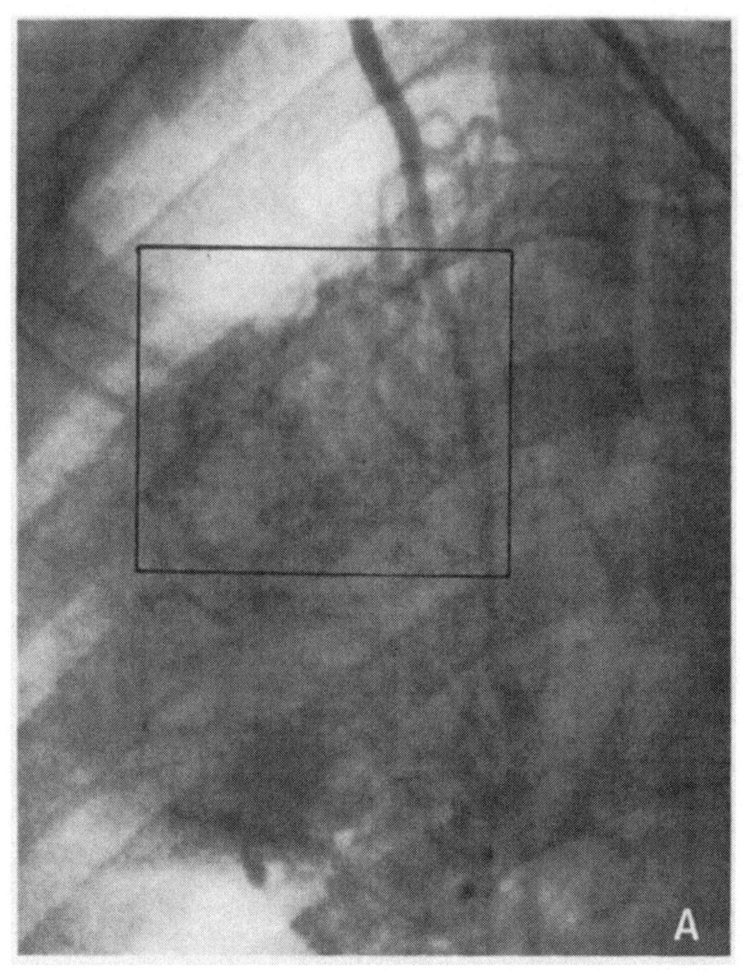


Fig. 57 A. Normal roentgenogram

## Case 12 (Chest): Mediastinal Tumor

Female, 16 yrs, student (Fig. 57 A and B).

*History:* Chest pain for previous 2 months, no complaint of cough, sputum.

Normal roentgenography reveals mass shadow at anterior mediasthinum. Mediastinal tumor was suspected and internal mammary normal arteriography and magnification arteriography was carried out. Exposure factors of magnification arteriography: 0.1 mm focal-spot tube, 120 kV, 3 mA, 0.1 sec, FFD 106 cm, focal spot-skin distance 25 cm. Chest thickness 16 cm.

Normal arteriography reveals that the tumor mass is imaged with the right internal mammary artery, which is tortuous and surrounds the tumor with a network of vessels (Fig. 57 A).

Magnification arteriography reveals the fine structure of vessels, which are enlarged or narrowed, with brush-like or scroll-like shadows. The arrangement of the vessels clearly shows lobulation (↗) (Fig. 57 B).

From these findings the origin of the tumor was considered to be the lymph node and malignant lymphoma was diagnosed. Surgical operation showed it to be HODGKIN's disease. The magnification made it possible to image the most peripheral portion of the vessels in the tumor.

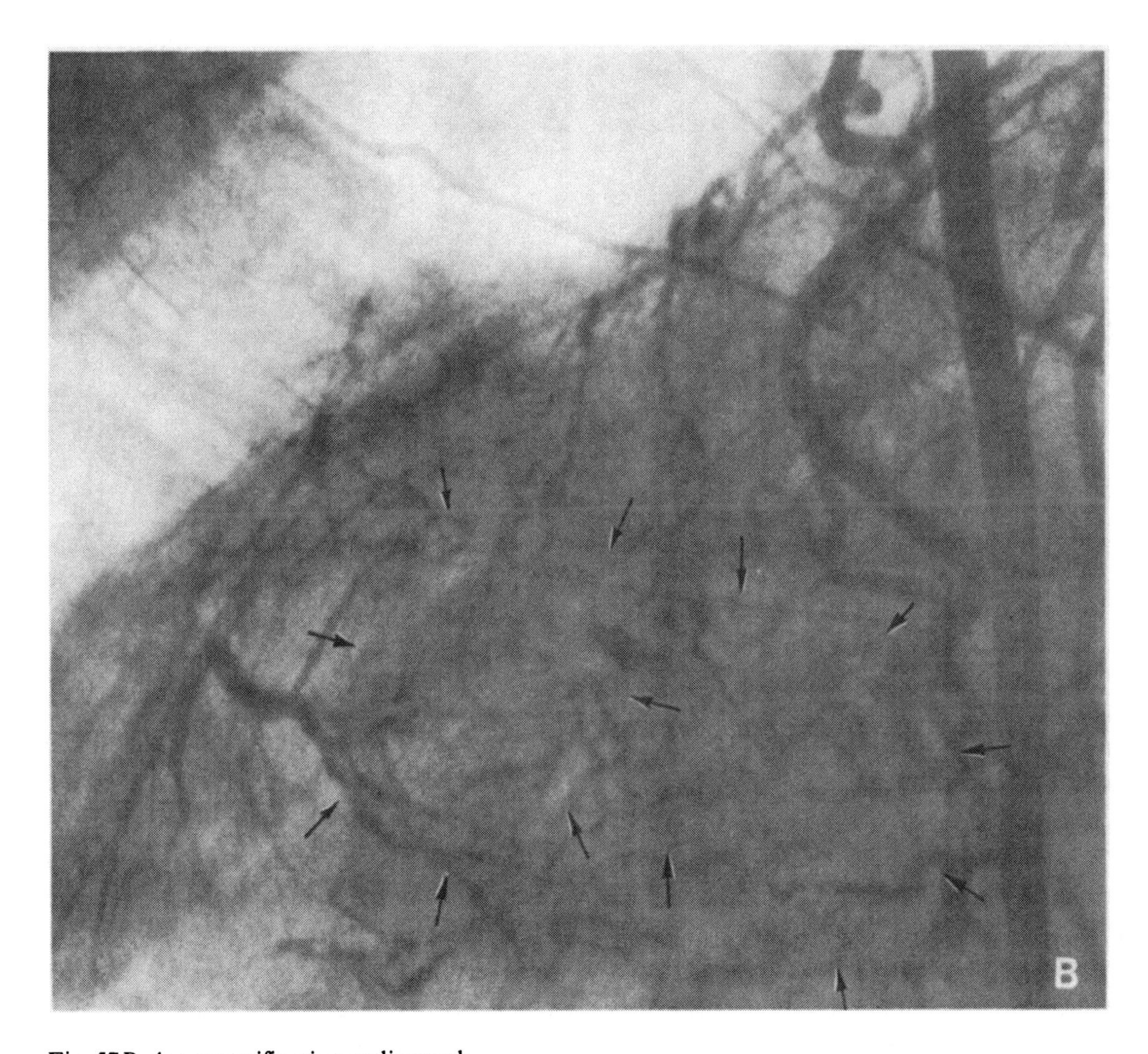


Fig. 57 B. 4 × magnification radiograph

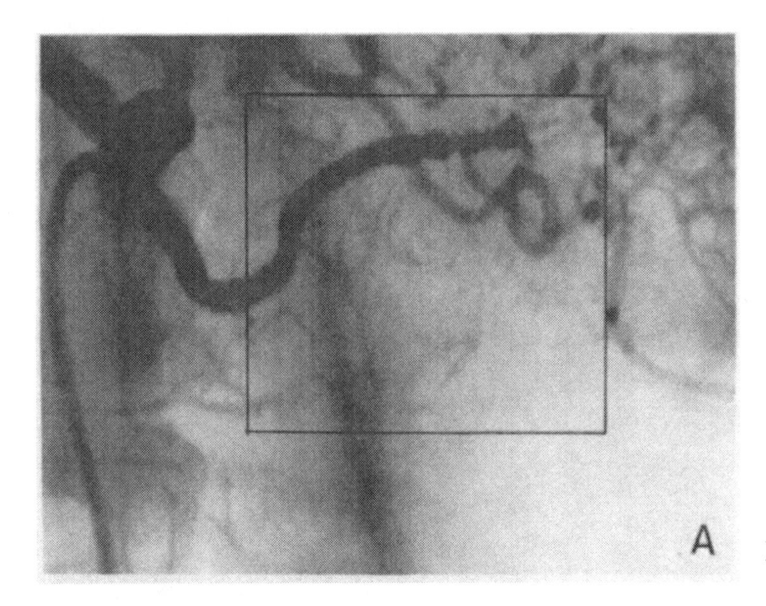


Fig. 58 A. Normal roentgenogram

## Case 13 (Abdomen): Carcinoma of Body and Tail of Pancreas

Male, 64 yrs, social worker (Fig. 58 A and B).

*History:* Epigastralgia for 4–5 months. No nausea or vomiting. Loss of appetite. No jaundice. X-ray examination of stomach and duodenum revealed pitting on the posterior wall, extending from the body of the stomach to the greater curvature. Endoscopic examination of the stomach revealed the gastric mucosa to be normal but a bulge was noted on the posterior wall of the body of the stomach.

Celiac arteriography was conducted by normal roentgenography and magnification radiography. Exposure conditions for 3 × magnification radiography: 0.1 mm focal-spot tube, 120 kV, 15 mA, 0.1 sec; FFD 120 cm. Abdominal thickness 19 cm.

Normal roentgenographic findings show no special changes in the branching of the A. coeliaca and no abnormality except for a permanent wave in the A. lienalis at the hilum of the spleen. Of the arteries of the pancreas, the path of the A. pancreatica transversus at the periphery is markedly tortuous and the splenic branches to the left of the body and tail of the pancreas are conspicuously bent. There is no finding of hypervascularity or tumor stain (Fig. 58 A).

The magnification arteriogram reveals distortion of the wall of the A. linealis around the branching of A. pancreatica magna. Also noted are interruption (↙) of the A. pancreatica transversus and irregularity (♂) of its wall. The A. pancreatica magna is markedly bent after branching (⤯) and its terminal branches show great irregularities in diameter and are arranged globularly (⤯). Along the lower border proliferation of small blood vessels is noted (Fig. 58 B).

From these findings a diagnosis of carcinoma of the body and tail of the pancreas was established. The small blood vessels that do not appear in normal arteriography are imaged by magnification radiography and confirm the presence of a tumor. This finding is significant.

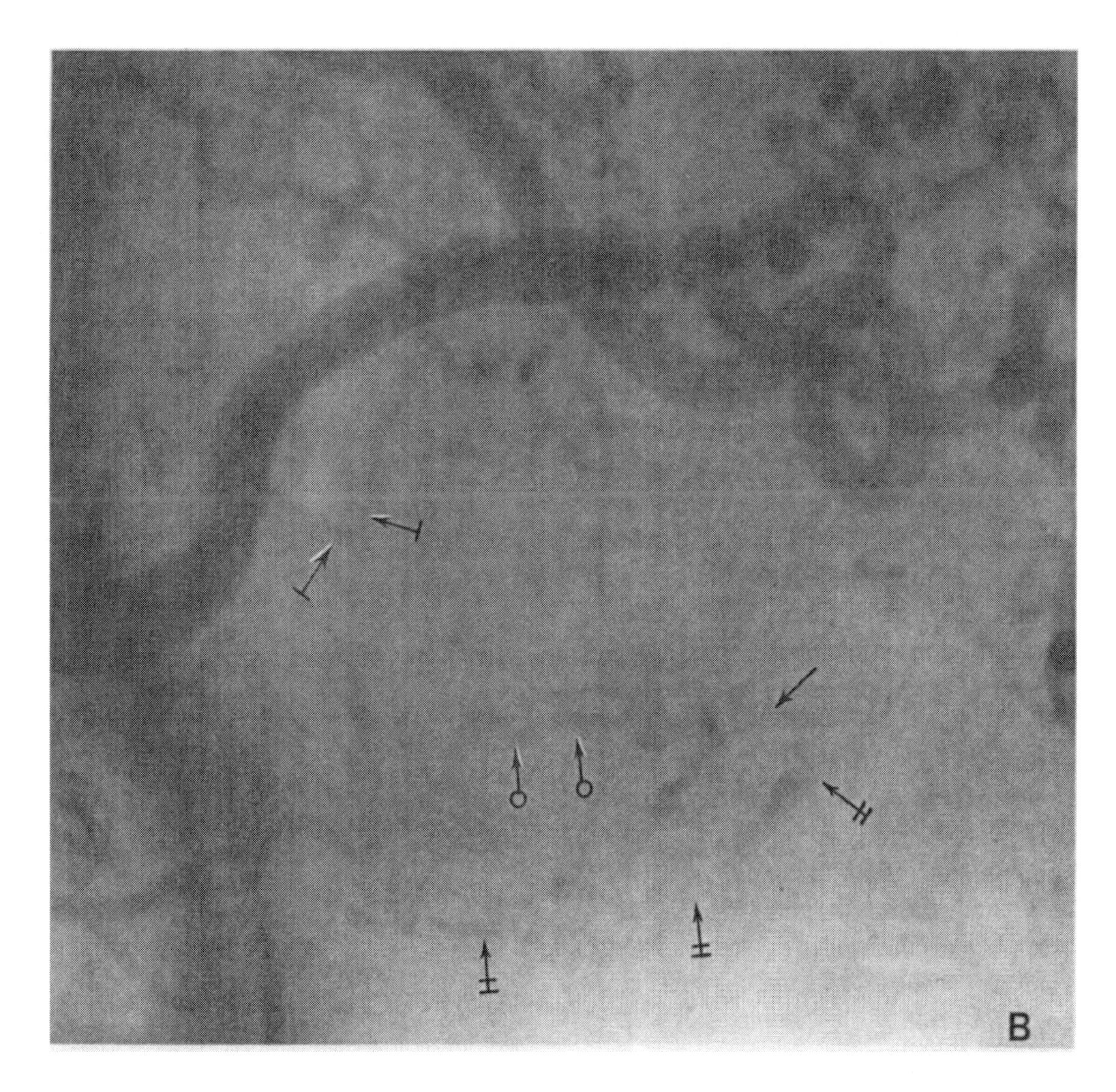


Fig. 58 B. 3 × magnification radiograph

## Case 14 (Abdomen): Renal Cell Carcinoma

Male, 57 yrs (Fig. 59 A and B).

*History:* No record of illness in his history. One week previously bloody urine was noticed. Neither abdominal pain nor pain on urination. Pyelogram reveals absence of image of calix at the upper pole of the kidney.

Normal and magnification selective renal arteriography were carried out. Exposure conditions of magnification radiography: 0.1 mm focal-spot tube, 110 kV, 15 mA, 0.1 sec. Abdomen thickness 21 cm.

Normal roentgenogram reveals an enlarged kidney with hypovascularity in the upper half. The image of A. arcuatae is elongated and mixed with blurred, flecked shadows (Fig. 59 A).

The magnification radiograph reveals tortuous and narrowed small vessels with a brush-like arrangement (×), seen as flecked shadows on normal roentgenogram. This is due to invasion by the tumor (Fig. 59 B).

The so-called tumor stain image in a normal angiogram often reveals such a fine structure on a magnification angiogram. The image of the glomerulus (↑) is also seen in the peripheral part of the kidney where there is no disease.

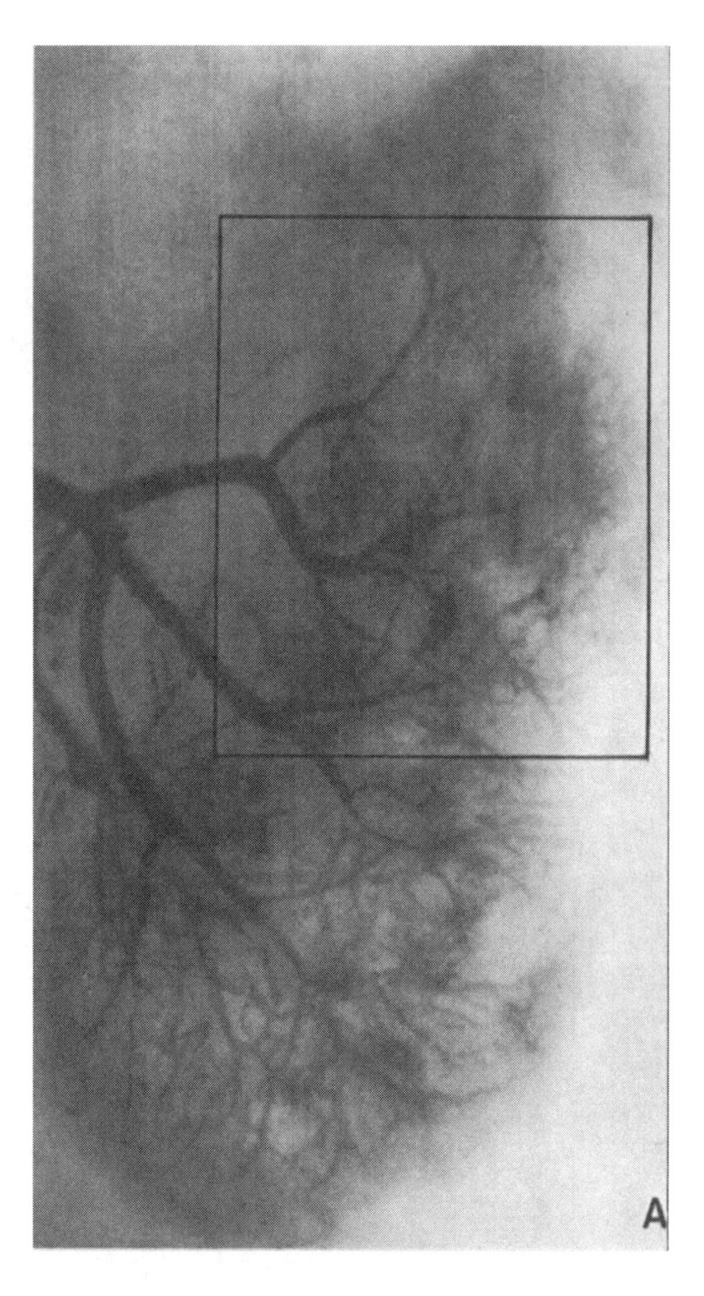


Fig. 59 A. Normal roentgenogram

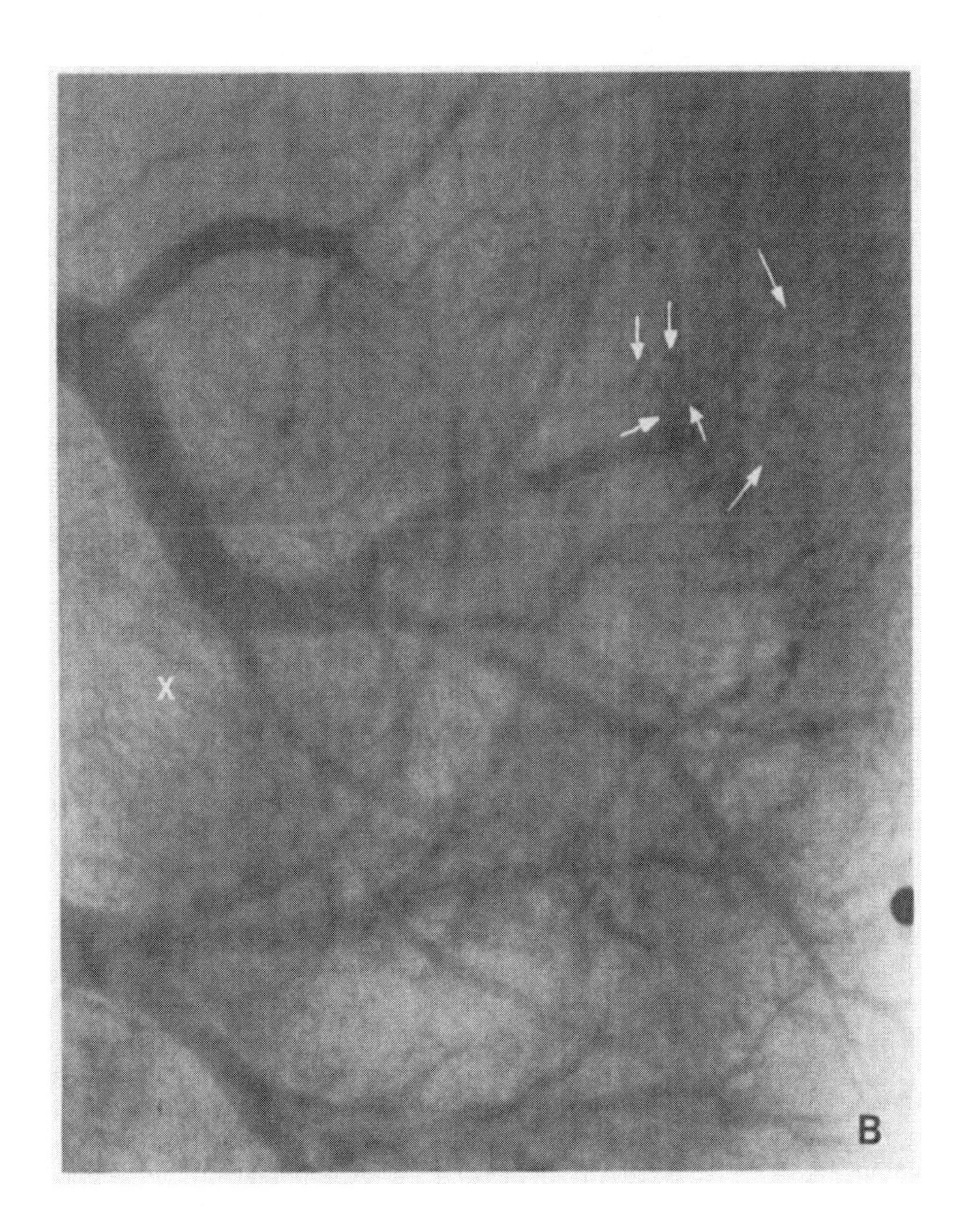


Fig. 59 B. 3 × magnification radiograph

# References


1. ABEL, M. S.: Advantages and limitations of the 0.3 mm focal spot tube for magnification and other technics. Radiology **66**, 747-752 (1955).
2. ABLOW, R.C., GREENSPAN, R.H., GLUCK, L.: The advantages of direct magnification technic in the newborn chest. Radiology **92**, 745-750 (1969).
3. ADERHOLD, K., SEIFERT, L.: Ergebnisse der radiologischen Vergrößerungstechnik mit einer neuen Feinstfokusröntgenröhre für Abbildungsmaßstäbe größer als 2:1. Fortschr. Röntgenstr. **81**, 181-194 (1954).
4. ADERHOLD, K.: Die erforderliche radiologische Vergrößerung bei Fokusdimensionen unter 0.3 mm. Acta radiol. (Stockh.) **43**, 329-341 (1955).
5. ALLEN, C., ALLEN, E. P.: Enlargement radiography with a 0.3 mm focus. Brit. J. Radiol. **26**, 474-480 (1953).
6. ANGERSTEIN, W., STARGARDT, A.: Grundlagen der direkten (geometrischen) Röntgenvergrößerung und der nachträglichen Vergrößerung von Röntgenbildern. Radiol. diagn. (Berl.) **12**, 280-288 (1971).
7. AYAKAWA, Y., SAKUMA, S., OKUMURA, Y.: Optimal magnification ratio for direct macroradiography. Studies with Modulation Transfer Function (MTF). Studies on enlargement radiography, 37th report. Nippon Acta radiol. **27**, 575-578 (1967) [in Japanese].
8. AYAKAWA, Y.: Modulation Transfer Function of films, intensifying screens and scanning spot size of microphotometer in direct fourfold macroradiography. Studies on enlargement radiography, 39th report. Nippon Acta radiol. **28**, 400-404 (1968) [in Japanese].
9. AYAKAWA, Y., SAKUMA, S.: Influence of scatterer on image deterioration and determination on optimal magnification ratio in direct macroradiography. Studies on enlargement radiography, 44th report. Nippon Acta radiol. **31**, 540-549 (1971) [in Japanese].
10. AYAKAWA, Y.: Optimal magnification ratio of direct macroradiography in high magnification. Modulation transfer function study on system combined with intensifying screen-film and object. Nagoya J. med. Sci. **34**, 227-240 (1972).
11. BAKER, C. D., LANE, F. E., PIRKEY, E. L.: Roentgen examination of old and new trauma of the spine with the ultra-fine focus roentgen tube. Amer. J. Roentgenol. **75**, 144-148 (1956).
12. BOOKSTEIN, J. J., VOEGELI, E.: A critical analysis of magnification radiography. Laboratory investigation. Radiology **98**, 23-30 (1971).
13. BOOKSTEIN, J. J., STECK, W.: Effective focal spot size. Radiology **98**, 31-33 (1971).
14. BOOKSTEIN, J. J., POWELL, T. J.: Short-target-film rotating-grid magnification. Comparison with air-gap magnification. Radiology **104**, 399-402 (1972).
15. BÜCHNER, H.: Direkte Röntgenvergrößerung und normale Aufnahme. Vergleichende Untersuchungen zur klinischen Abgrenzung (1. Teil). Fortschr. Röntgenstr. **80**, 71-87 (1954).
16. BÜCHNER, H.: Direkte Röntgenvergrößerung und normale Aufnahme. Vergleichende Untersuchungen zur klinischen Abgrenzung (2.Teil). Fortschr. Röntgenstr. **80**, 502-514 (1954).
17. BÜCHNER, H.: Die Indikation zur direkten Röntgenvergrößerung bei Knochenaufnahmen. Radiologe **1**, 222-229 (1961).
18. BURGER, G.C.E., COMBEE, B., VAN DER TUUK, J.H.: X-Ray fluoroscopy with enlarged image. Philips Tech. Rev. **8**, 321-329 (1946).

19. CALENOFF, L., NORFRAY, J.: Magnification digital roentgenography: A method for evaluating renal osteodystrophy in hemodialyzed patients. Amer. J. Roentgenol. **118**, 282–292 (1973).
20. COLTMAN, J. W.: The specification of imaging properties by response to a sine wave input. J. opt. Soc. Amer. **44**, 468–471 (1954).
21. COMBEÉ, B., BOTDEN, P. J. M.: Special X-ray tubes. Philips Tech. Rev. **13**, 71–80 (1951).
22. CREMIN, B. J.: Macroradiography in severe neonatal heart disease. Brit. J. Radiol. **45**, 75 (1972).
23. DOI, K.: Optical transfer functions of the focal spot of X-ray tubes. Amer. J. Roentgenol. **94**, 712–718 (1965).
24. DOI, K., ROSSMANN, K.: Computer simulation of small blood vessel imaging in magnification radiography. In: Small vessel angiography—imaging, morphology, physiology and clinical applications (ed. S. K. HILAL), p. 6-12. Mosby: Saint Louis, 1973.
25. DOI, K., ROSSMANN, K.: Longitudinal magnification in radiologic images of thick objects: A new concept in magnification radiography. Radiology **114**, 443–447 (1975).
26. DÜNISCH, O., PFEILER, M., KUHN, H.: Problems and aspects of the radiological magnification technique. Electromedica **3**, 114–120 (1971).
27. FERRANT, W., SAN NICOLÓ, M. R.: Die förderliche Röntgenvergrößerung. Fortschr. Röntgenstr. **81**, 194–205 (1954).
28. FLETCHER, D. E., ROWLEY, K. A.: Radiographic enlargements in diagnostic radiology. Brit. J. Radiol. **24**, 598–604 (1951).
29. FRIEDMAN, P. J., GREENSPAN, R. H.: Observations on magnification radiography. Visualization of small blood vessels and determination of focal spot size. Radiology **92**, 549–557 (1969).
30. FUJITA, T.: X-ray television macrofluorography using a small focus tube. Consideration on information content and patient dose. Nippon Acta radiol. **30**, 615–625 (1970) [in Japanese].
31. GILARDONI, A., SCHWARZ, G. S.: Magnification of radiographic images in clinical roentgenology and its present-day limit. Radiology **95**, 866–878 (1952).
32. GREENSPAN, R. H., SIMON, A. L., RICKETTS, H. J., ROJAS, R. H., WATSON, J. C.: In vivo magnification angiography. Invest. Radiol. **2**, 419–431 (1967).
33. HACKER, H.: Detail angiography. Direct X-ray enlargement in serial angiography. Electromedica **6**, 346–349 (1970).
34. HALE, J., MISHKIN, M. M.: Serial direct magnification cerebral angiography. Amer. J. Roentgenol. **107**, 616–621 (1969).
35. HILAL, S. K.: Small vessel angiography—imaging, morphology, physiology, and clinical applications. Saint Louis: Mosby 1973.
36. HOLLANDER, B. A., HILAL, S. K., SEAMAN, W. B.: Evaluation of a radiographic imaging system with a microfocal spot X-ray tube. The usefulness of three- to fivefold magnification. Radiology **103**, 667–674 (1972).
37. IKEDA, H.: Stereomacroradiography in fourfold magnification. Studies on macroroentgenography, 46th report. Nippon Acta radiol. **31**, 895–907 (1971) [in Japanese].
38. IMAGUNBAI, N.: Macrolymphography in fourfold magnification. Studies on macroroentgenography, 47th report. Nippon Acta radiol. **31**, 785–800 (1971) [in Japanese].
39. ISHIGAKI, T.: First metatarsal-phalangeal joint of gout. Macroradiographic examination in 6 times magnification. Nippon Acta radiol. **33**, 839–854 (1973) [in Japanese].
40. KANEKO, M.: Percutaneous selective catheterization for enlargement angiography. Studies on enlargement angiography, first report. Studies on enlargement radiography, 31st report. Nippon Acta radiol. **24**, 479–484 (1964) [in Japanese].
41. KANEKO, M.: Angiography in four times magnification applied to the kidney of the living human body. Studies on enlargement angiography, 2nd report. Studies on enlargement radiography, 35th report. Nippon Acta radiol. **26**, 55–65 (1966) [in Japanese].
42. KANEKO, M., SAKUMA, S.: Renal glomerulography in fourfold magnification. Nagoya J. med. Sci. **34**, 297–302 (1972).
43. KEATS, T. E., KOENIG, G. F.: Magnification technic in roentgenography of the chest. Radiology **69**, 745–747 (1957).
44. KOMIYAMA, K.: Size of the focal spot of an auto-bias X-ray tube. Studies on enlargement radiography, 7th report. Nippon Acta radiol. **14**, 487–494 (1954) [in Japanese].

45. KOMIYAMA, K.: What type of intensifying screen is the most suitable for obtaining good X-ray image at the clinical use of enlargement radiography? Studies on enlargement radiography, 10th report. Nippon Acta radiol. **15**, 81–86 (1955) [in Japanese].
46. KOMIYAMA, K., FUJIMAKI, M., HASHIMOTO, Y., SAKAGAMI, K.: Pulmonary tuberculosis in enlargement radiography. Studies on enlargement radiography, 12th report. Nippon Acta radiol. **15**, 1028–1037 (1956) [in Japanese].
47. LALLEMAND, D., SAUVEGRAIN, J., MARESCHAL, J. L.: Laryngo-tracheal lesions in infants and children. Detection and follow-up studies using direct radiographic magnification. Ann. Radiol. **16**, 293–304 (1973).
48. LLOYD, G. A. S., JONES, B. R., WELHAM, R. A. N.: Intubation macrodacryocystography. Brit. J. Ophthal. **56**, 600–603 (1972).
49. MAEKOSHI, H., FUJITA, T., SAKUMA, S.: Skin dose and volume dose in macroradiography of high magnification. Nippon Acta radiol. **33**, 336–343 (1973) [in Japanese].
50. MATSUDA, T.: Evaluation of the direct enlargement radiography applied to the examination of the bone structure of adults. Studies on enlargement radiography, 8th report. Nippon Acta radiol. **14**, 767–774 (1955) [in Japanese].
51. MILNE, E.: Comments on radiographic magnification and on the design of cathodes for X-ray tubes. In: Small vessel Angiography—imaging, morphology, physiology and clinical applications by S. K. HILAL, p. 29–35. Mosby: Saint Louis 1973.
52. MOORE, R., KRAUSE, D., AMPLATZ, K.: A flexible grid-air gap magnification technique. Radiology **104**, 403–407 (1972).
53. MORGAN, R. H.: The frequency response function. A valuable means of expressing the informational recording capability of diagnostic X-ray systems. Amer. J. Roentgenol. **88**, 175–186 (1962).
54. NAKAYAMA, N., WENDE, S., SCHINDLER, E.: The clinical significance of the tentorial artery after operative treatment of congenital arteriosinus fistulas. Neuroradiology **6**, 196–199 (1973).
55. NEMET, A., COX, W. F.: The improvement of definition by X-ray image magnification. Brit. J. Radiol. **29**, 335–337 (1956).
56. NIXON, W. C.: Improved definition in X-ray diagnosis by high-voltage projection microradiography. Brit. J. Radiol. **29**, 657–662 (1956).
57. OKUMURA, Y., AYAKAWA, Y., SAKUMA, S.: Modulation Transfer Function of the very small focal spot of X-ray tube for macroradiography. Studies on enlargement radiography, 38th report. Nippon Acta radiol. **27**, 590–594 (1967) [in Japanese].
58. ONO, Y.: A study on the clinical use of direct enlargement radiography for diagnosis of an early stage silicosis. Jap. J. clin. Tuberc. **17**, 55–60 (1958) [in Japanese].
59. ŌNO, A., ISHIGAKI T., ORITO, T.: The method of directing the central X-ray beam to the small focus in the body. Sakura X-ray photo. Rev. **107**, 16–29 (1974) [in Japanese].
60. PFEILER, M., THEIL, G.: On the radiographic magnification technique in lymphography. Electromedica **3**, 231–234 (1970).
61. PLAATS, G. J. VAN DER: Prinzipien, Technik und medizinische Anwendung der radiologischen Vergrößerungstechnik. Fortschr. Röntgenstr. **77**, 605–610 (1952).
62. RANDALL, P. A., AMPLATZ, K.: Magnification coronary arteriography. Radiology **101**, 51–56 (1971).
63. RAO, G. U. V., CLARK, R. L.: Radiographic magnification versus optical magnification. Radiology **94**, 196 (1970).
64. RAO, G. U. V.: A new method to determine the focal spot size of X-ray tubes. Amer. J. Roentgenol. **111**, 628–633 (1971).
65. RAO, G. U. V., SOONG, A.-L.: Physical characteristics of modern microfocus X-ray tubes. Amer. J. Roentgenol. **119**, 626–634 (1973).
66. RUPP, N., RADKE, J.: Knochenstrukturanalyse durch Vergrößerung im Röntgenbild. Fortschr. Röntgenstr. **118**, 335–339 (1973).
67. SAKUMA, S.: Normal lung markings imaged by radiography in direct 4 times magnification. Studies on enlargement radiography, 26th report. Nippon Acta radiol. **20**, 368–377 (1960) [in Japanese].
68. SAKUMA, S., KOGA, S.: Small intestinal mucosal patterns imaged by radiography in direct 4 times magnification. Studies on enlargement radiography, 29th report. Nippon Acta radiol. **21**, 627–633 (1961) [in Japanese].

69. SAKUMA, S.: Angiography in direct 4 times magnification applied to field of ophthalmology. Tohoku J. exp. Med. **82**, 242–249 (1964).
70. SAKUMA, S., HAYASHI, F., KOGA, S., HIRAMATSU, K., IMAGUNBAI, N.: Macrographic examination of pulmonary stippling shadow following lymphography. Studies on enlargement radiography, 34th report. Nippon Acta radiol. **25**, 1419–1424 (1966) [in Japanese].
71. SAKUMA, S., KOGA, S., IMAGUNBAI, N., IKEDA, H., AYAKAWA, Y.: Filling procedure with contrast media in lymph nodes in macroradiography. Clin. Radiol. (Tokyo) **11**, 237–242 (1966) [in Japanese].
72. SAKUMA, S., HAYASHI, F., TAKEUCHI, A., AYAKAWA, Y.: Findings of normal roentgenogram with comparison to that of macroradiogram and axial transverse tomogram. Clin. Radiol. (Tokyo) **12**, 665–672 (1967) [in Japanese].
73. SAKUMA, S., HAYASHI, F., KOGA, S., HIRAMATSU, K., IMAGUNBAI, N.: Über die vierfache direkte Vergrößerungsaufnahme von Lungenkomplikationen nach der Lymphographie. Radiologe **8**, 221–224 (1968).
74. SAKUMA, S., KOGA, S., IMAGUNBAI, N., IKEDA, H., AYAKAWA, Y.: Macroradiography in four times magnification applied to serial lymphography. Radiologe **8**, 224–228 (1968).
75. SAKUMA, S., MIURA, T.: Selective pulmonary arteriography in direct 4 times magnification. Studies on macroroentgenography, 40th report. Nippon Acta radiol. **28**, 1283–1287 (1968) [in Japanese].
76. SAKUMA, S., KATO, T., AYAKAWA, Y.: Hypotonic gastroduodenal macroradiography. Clin. Radiol. (Tokyo) **13**, 543–547 (1968) [in Japanese].
77. SAKUMA, S., IKEDA, H., AYAKAWA, Y., TANAKA, Y., TAKAHASHI, S.: Angiography with direct fourfold magnification. Invest. Radiol. **4**, 310–316 (1969).
78. SAKUMA, S., AYAKAWA, Y., OKUMURA, Y., MAEKOSHI, H.: Determination of focal-spot characteristics of microfocus X-ray tubes. Invest. Radiol. **4**, 335–339 (1969).
79. SAKUMA, S., AYAKAWA, Y., FUJITA, T.: Macroroentgenography in twentyfold magnification taken by means of 50 μ focal spot X-ray tube and evaluation of its reduced image. Studies on enlargement radiography, 42nd report. Nippon Acta radiol. **30**, 205–209 (1970) [in Japanese].
80. SAKUMA, S., AYAKAWA, Y., MAEKOSHI, H., FUJITA, T., ISHIGAKI, T., OHARA, K.: Small field macroradiography in high magnification. Studies on enlargement radiography, 48th report. Nippon Acta radiol. **31**, 1115–1121 (1972) [in Japanese].
81. SAKUMA, S.: Clinical evaluation of direct magnification radiography. Sakura X-ray photo. Rev. **105**, 7–12 (1974) [in Japanese].
82. SAMUEL, E.: Macroradiography (enlargement technique) in the radiology of the ear, nose and throat. Brit. J. Radiol. **26**, 558–567 (1953).
83. SANDOR, T., ADAMS, D. F.: Computer model to study image formation of blood vessels smaller than the exposing focal spot. Radiology **109**, 195–200 (1973).
84. SANDOR, T., ADAMS, D. F., HERMAN, P. G., EISENBERG, H., ABRAMS, H. L.: The potential of magnification angiography. Amer. J. Roentgenol. **120**, 916–921 (1974).
85. SASAKI, T., KIDO, C., SAKUMA, S.: Macroradiography in 4 times magnification applied to cerebral angiography. Nagoya J. med. Sci. **29**, 245–249 (1967).
86. SASAKI, T.: Direct four-fold bronchial macroangiography in inflammatory and neoplastic diseases. Gann Monograph No. 9, 47–54 (1970).
87. SATO, N.: Bronchial arteriography in fourfold magnification. Studies on enlargement angiography, 3rd report. Studies on enlargement radiography, 42nd report. Nippon Acta radiol. **29**, 454–459 (1969) [in Japanese].
88. SAGEL, S. S., GREENSPAN, R. H.: Minute pulmonary arteriovenous fistulas demonstrated by magnification pulmonary angiography. Radiology **97**, 529–530 (1970).
89. SEIFERT, L.: Entwicklung einer Feinstfokus-Röntgenröhre mit elektrostatischer Fokussierung und Untersuchungen über die spezifische Anodenbelastbarkeit bei sehr kleinen Brennflecken. Exp. Techn. d. Physik **2**, 109–126, 154–170 (1954).
90. SEYSS, R.: Die Strukturzeichnung der peripheren Lungenabschnitte auf der direkten Vergrößerungsaufnahme. Fortschr. Röntgenstr. **81**, 32–35 (1954).

91. SHINOZAKI, T., KOMIYAMA, K.: Enlargement radiography of silicosis. Studies on enlargement radiography, 16th report. Nippon Acta radiol. **17**, 957–962 (1957) [in Japanese].
92. SHINOZAKI, T., KOMIYAMA, K., YOSHIDA, M., OKAJIMA, S.: Effect of superposition for image formation at enlargement radiography. Studies on enlargement radiography, 19th report. Nippon Acta radiol. **18**, 324–328 (1958) [in Japanese].
93. SIDAWAY, M. E.: Small-vessel changes in renal disease. Brit. med. Bull. **28**, 247–249 (1972).
94. SPIEGLER, G.: Physikalische Grundlagen der Röntgendiagnostik, S. 47. Stuttgart: Thieme 1957.
95. STARGARDT, A.: Vergleich der geometrischen und elektronenoptischen Vergrößerung von Röntgenbildern aus strahlenphysikalischer Sicht. Radiol. diagn. (Berl.) **12**, 288–295 (1971).
96. STEIN, H. L.: Direct serial magnification renal arteriography: A clinical study. J. Urol. (Baltimore) **109**, 967–970 (1973).
97. SUGIE, Y.: Study on radiograms of pulmonary silicosis taken by means of radiography in 4 times magnification. Studies on enlargement radiography, 27th report. Nippon Acta radiol. **19**, 2077–2089 (1960) [in Japanese].
98. SUGIURA, Y.: Clinical application of enlargement radiography in orthopaedic surgery (Parts 1, 2, 3). Nagoya J. med. Sci. **21**, 333–338 (1958); **22**, 1–14 (1958); **22**, 131–143 (1959).
99. SVOBODA, M.: Extraartikuläre Ankylose beider Hüftgelenke beim Bluter. Zbl. Chir. **82**, 873–874 (1957).
100. SVOBODA, M., MALY, VL.: Les altérations articulaires dans l'hémophilie. Images radiologiques, tomographiques et agrandies directement. J. Radiol. Électrol. **39**, 610–617 (1958).
101. TAKAHASHI, M., NAGATA, Y.: Angiography of trophoblastic tumors. A correlation of pelvic angiography with direct four-fold magnification angiography of uterine specimens. Amer. J. Roentgenol. **112**, 779–787 (1971).
102. TAKAHASHI, S., KOMIYAMA, K.: Radioactive isotope $S^{35}$ as a radiation source in microradiography. Studies on enlargement radiography, first report. Hirosaki med. J. **3**, 27–29 (1952) [in Japanese].
103. TAKAHASHI, S., KOMIYAMA, K.: Direct microradiography taken with hard X-rays. Studies on enlargement radiography, 2nd report. Hirosaki med. J. **3**, 148–153 (1952) [in Japanese].
104. TAKAHASHI, S.: Enlargement radiography. Fuji X-ray Jiho Suppl. (1954) [in Japanese].
105. TAKAHASHI, S., KOMIYAMA, K.: Radiography with very fine focus of self-bias principle. Studies on enlargement radiography, 5th report. Nippon Acta radiol. **14**, 220–226 (1954) [in Japanese].
106. TAKAHASHI, S., KUBOTA, Y., YOSHIDA, M.: Über die Vergrößerung des Querschnittbildes des Körpers mittels Röntgenstrahlen. Fortschr. Röntgenstr. **80**, 387–392 (1954).
107. TAKAHASHI, S., KOMIYAMA, K., TANAKA, M.: A fixed anode tube with a very fine focus made with autobiased electron beam. Its application to enlargement radiography. Studies on enlargement radiography, first report. Tohoku J. exp. Med. **62**, 253–259 (1955).
108. TAKAHASHI, S., WATANABE, T., TANAKA, M.: Ultrasmall focal spot tube applied to clinical use. Studies on enlargement radiography, 13th report. Nippon Acta radiol. **15**, 838–842 (1955) [in Japanese],
109. TAKAHASHI, S., YOSHIDA, M.: Roentgenography in high magnification. Reliability and limitation of enlargement radiography. Acta radiol. (Stockh.) **48**, 280–288 (1957).
110, TAKAHASHI, S., WATANABE, T.: Rotatory anode X-ray tube with biased small focal spot applied to clinical practice. Studies on enlargement radiography, 15th report. Nippon Acta radiol. **17**, 77–80 (1957) [in Japanese].
111. TAKAHASHI, S., YOSHIDA, M.: Radiographic effects influencing image formation in high magnification. Direct enlargement radiography study, 3rd report. Nagoya J. med. Sci. **21**, 115–127 (1958).
112. TAKAHASHI, S., SAKUMA, S., SUGIE, Y.: Vierfache direkte Vergrößerungsaufnahmen der Lungen bei gesunden und bei frühen silikotischen Personen. Fortschr. Röntgenstr. **92**, 294–301 (1960).

113. TAKAHASHI, S., HASHIMOTO, Y., FUKUTA, M.: Direct enlargement arteriography of peripheral vessels. In: Vascular roentgenology—arteriography, phlebography, lymphography (ed. R. A. SCHOBINGER and F. F. RUZICKA, Jr., p. 465-468. New York: Macmillan 1964.

114. TAKAHASHI, S., SAKUMA, S., KANEKO, M., KOGA, S.: Angiography at fourfold magnification with special reference to the examination of tumours. Acta radiol. diagn. **4**, 206-216 (1966).

115. TAKAHASHI, S., SAKUMA, S., AYAKAWA, Y.: Die vierfache direkte Vergrößerungsaufnahme. Radiologe **8**, 217-221 (1968).

116. TAKAHASHI, S., SASAKI, T., SAKUMA, S., TOBITA, K.: X-ray television macrofluoroscopy. In: Television in diagnostic radiology (ed. R. D. MOSELEY, Jr., and J. H. RUST), p. 121-140. Birmingham, Ala.: Aesculapius 1969.

117. TAKAHASHI, S.: An atlas of axial transverse tomography and its clinical application. Berlin-Heidelberg-New York: Springer 1969.

118. TAKAHASHI, S., SAKUMA, S., WATANABE, T.: Makroradiographie mit hoher Vergrößerung. Radiol. diagn. (Berl.) **12**, 300-312 (1971).

119. TAKAHASHI, S., SAKUMA, S.: Roentgenography and roentgen examination. Tokyo: Nanzando 1972 [in Japanese].

120. TAKAHASHI, S., SAKUMA, S., AYAKAWA, Y., MAEKOSHI, H., OHARA, K.: Radiation levels of macroradiography: Radiation exposure and image quality. Radiology **112**, 709-713 (1974).

121. TAKARO, T., SCOTT, S. M.: Angiography using direct roentgenographic magnification in man. Amer. J. Roentgenol. **91**, 448-452 (1964).

122. TAKARO, T.: Experimental renal glomerulography. Amer. J. Roentgenol. **101**, 681-687 (1967).

123. TAMAKI, H.: Macroradiography of all teeth and periodontal structures. Proc. 3rd International Congress of Maxillofacial Radiology. April 18-21, 1974, Kyoto, Japan.

124. TOKUNAGA, O.: Direct 4 times enlargement radiography applied to pulmonary tuberculosis. Studies on enlargement radiography, 28th report. Nippon Acta radiol. **19**, 2315-2330 (1960) [in Japanese].

125. VALLEBONA, A.: Radiography with great enlargement (microradiography) and a technical method for the radiographic dissociation of the shadow. Radiology **17**, 340-341 (1931).

126. WACKENHEIM, A., NAKAYAMA, N., WENDE, S.: Magnification of the veins in vertebral angiography. Neuroradiology **6**, 56-59 (1973).

127. WATANABE, T.: Contrast medium suitable for application to enlargement radiography. Studies on enlargement radiography, 17th report. Nippon Acta radiol. **17**, 1427-1431 (1958) [in Japanese].

128. WATANABE, T.: Three times magnification radiography of stomach. Clin. Radiol. (Tokyo) **16**, 11-16 (1971) [in Japanese].

129. WENDE, S., SCHINDLER, K.: Technique and use of X-ray magnification in cerebral arteriography. Neuroradiology **1**, 117-120 (1970).

130. WENDE, S., SCHINDLER, K., MORITZ, G.: Der diagnostische Wert der angiographischen Vergrößerungstechnik mit Feinst-Fokus-Röhren in 2 Ebenen. Radiologe **11**, 471-475 (1971).

131. WENDE, S., NAKAYAMA, N.: Magnification angiography in orbital diseases. Neuroradiology **5**, 187-189 (1973).

132. WENDE, S., NAKAYAMA, N., PALVÖLGYI, R., SCHINDLER, K.: Comparison of cerebral scintigraphy with magnification angiography under hyperventilation in cerebral tumors. Neuroradiology **5**, 78-82 (1973).

133. WENDE, S., Zieler, E., NAKAYAMA, N.: Cerebral magnification angiography-physical basis and clinical results. Berlin-Heidelberg-New York: Springer 1974.

134. WERNER, K., BADER, W.: Über die röntgenologische Erfassung kleiner Knochendefekte durch direkte Röntgenvergrößerung und Vergrößerungstomographie mit Feinstfokusröhren. Fortschr. Röntgenstr. **80**, 87-101 (1954).

135. WOOD, E. H.: Preliminary observations regarding value of a very fine focus tube in radiologic diagnosis. Radiology **61**, 382-389 (1953).

136. YATO, M.: Evaluation of various kinds of effects on the macroradiography. Nippon

Acta radiol. **18**, 1595-1602 (1959) [in Japanese].

137. YATO, M.: Distortion effect in macroradiography. Nippon Acta radiol. **19**, 252-255 (1959) [in Japanese].
138. YOSHIDA, M.: Lung markings imaged by enlargement radiography. Studies on enlargement radiography, 11th report. Nippon Acta radiol. **15**, 91-99 (1955) [in Japanese].
139. YOSHIDA, M.: Fundamental study on the direct radiography in high magnification. Studies on enlargement radiography, 16th report. Nippon Acta radiol. **17**, 1418-1426 (1958) [in Japanese].
140. ZIMMER, E. A.: Methodische Bemerkungen und Leitsätze zur direkten Röntgen-Vergrößerung. Fortschr. Röntgenstr. **75**, 292-302 (1951).
141. ZIMMER, E. A.: Die praktische Anwendung und die Ergebnisse der radiologischen Vergrößerungstechnik. Fortschr. Röntgenstr. **78**, 164-169 (1953).

# Subject Index

CPSIA information can be obtained at www.ICGtesting.com
Printed in the USA
LVOW070308211112
308292LV00001B/11/P

9 783642 661228